Loss of Control of Cesium-137 Well Logging Source Resulting in Radiation Exposures to Members of the Public

U.S. Nuclear Regulatory Commission
Region IV
Arlington, TX 76011-4005

AVAILABILITY OF REFERENCE MATERIALS
IN NRC PUBLICATIONS

Loss of Control of Cesium-137 Well Logging Source Resulting in Radiation Exposures to Members of the Public

Manuscript Completed: April 2004
Date Published: October 2004

Prepared by
D. Boal, R.Brown, R. Leonardi,
M. Shaffer, S. Sherbini

U.S. Nuclear Regulatory Commission
Region IV
611 Ryan Plaza Drive
Arlington, TX 76011-4005

ABSTRACT

This report describes the events that occurred on a drill rig in Montana on May 21, 2002, that led to the unplanned radiation exposure of 31 rig workers. These workers were not radiation workers, and were therefore considered by the Nuclear Regulatory Commission to be subject to the Agency's dose limit for members of the public, which is 0.1 centisievert (cSv) (0.1 rem) per year. The doses assessed by the Nuclear Regulatory Commission to have been received by the workers as a result of this incident were, for most of the exposed workers, above the dose limit for members of the public, but in all cases was far below the dose limit for radiation workers of 5 cSv (5 rem) per year. Although the doses received are relatively low and are not expected to cause any clinical effects, they are in violation of the Nuclear Regulatory Commission's regulations.

Included in the report is a detailed description of the sequence of events, the root and other causes for the event, a detailed description of the methods used to assess the worker's doses, and a discussion of the biological dosimetry undertaken for some of the exposed workers to support and verify the dose assessments. A discussion of the less than adequate state of biological dosimetry in the United States is also included.

CONTENTS

Appendices

Figures

Tables

TEAM MEMBERS

The following are the NRC Regional and Headquarters staff who were on the Augmented Inspection Team.

Dennis Boal, OI, Region IV Field Office

Robert Brown, Region IV

Richard Leonardi, Region IV

Mark Shaffer, Team Leader, Region IV

Sami Sherbini, NMSS, Headquarters

ACKNOWLEDGMENTS

The team wishes to acknowledge the technical assistance provided by the following.

William Blakely, Ph.D., Armed Forces Radiobiological Research Institute, Bethesda, MD

Susanne Gollin, Ph.D., University of Pittsburgh Cancer Institute, Pittsburgh, PA

John Hunt, Ph.D., *Comissão National de Energia Nuclear*, Rio de Janeiro, Brazil

David Lloyd, Ph.D., National Radiological Protection Board, Chilton, Didcot, Oxon, UK

Daniel Lundstrom, Precision Drilling, Nisku, Alberta, Canada

Pat Prasanna, Ph.D., Armed Forces Radiobiological Research Institute, Bethesda, Maryland

Adriana Ramalho, Ph.D. *Comissão National de Energia Nuclear*, Rio de Janeiro, Brazil

Robert Ricks, M.D., Radiation Emergency Assistance Center/Training Site, Oak Ridge, TN

Neil Wald, M.D., University of Pittsburgh, Pittsburgh, PA

In addition to the individuals noted above, the AIT wishes to thank the many individuals who provided freely of their time, knowledge, and expertise to help the team complete its work.

ABBREVIATIONS

AFRRI	Armed Forces Radiobiological Research Institute
AIT	Augmented Inspection Team
CDT	Central Daylight Time
CFR	*Code of Federal Regulations*
cGy	centigray
Ci	curie
cSv	centisievert
DOE	Department of Energy
gBq	gigabequerel
Gy	gray
IMC	Inspection Manual Chapter
IP	Inspection Procedure
IRD	*Instituto de Radioprotecao e Dosimetria*
ISO	International Organization for Standardization
LCM	Loss Causation Model
LE	Logging Engineer
MCNP	Monte Carlo N-Particle
MORT	Management Oversight and Risk Tree
NMSS	NRC Office of Nuclear Material Safety and Safeguards
NRC	U.S. Nuclear Regulatory Commission
NRPB	National Radiological Protection Board
NSSI	Nuclear Source and Services, Inc.
OI	NRC Office of Investigations
ORAU	Oak Ridge Associated Universities
ORISE	Oak Ridge Institute for Science and Education
REAC/TS	Radiological Emergency Assistance Center/Training Site
RSO	Radiation Safety Officer
SOS	Schlumberger Oilfield Services
STC	Schlumberger Technology Corporation
TI	Temporary Instruction

1 INTRODUCTION

1.1 Overview

This report describes the activities undertaken by the U.S. Nuclear Regulatory Commission's (NRC) Augmented Inspection Team (AIT) in connection with an incident involving the exposure of a number of members of the public to radiation doses that in some cases exceeded the NRC's dose limit for members of the public. The incident occurred on a drill rig in the State of Montana on May 21st, 2002, and was reported to the NRC on May 23rd, 2002. NRC's initial response to the notification was to send a reactive inspection team to the site to determine the details of the incident. However, a blood test (cytogenetics) performed on one of the workers suggested that the worker had been exposed to a radiation dose of the order of 2 gray (Gy) (200 rad), the inspection was upgraded to an AIT.

1.2 Use of Byproduct Material in Oil and Gas Well Logging

Well logging companies use instruments lowered into a hole drilled in the ground, called a well, to obtain information about underground rock formations, such as type of rock, porosity, density, and hydro-carbon content. The instruments are lowered into the well, which may be from a few hundred to 30,000 feet deep, on a cable known as a wireline. The wireline carries the signals from the logging instruments to the surface for analysis. Information collected in this manner is recorded and plotted on charts as the logging instruments are slowly raised from the bottom of the well. This information is studied and interpreted by experienced geologists or engineers to determine the likely presence and amounts of oil or gas. Sealed radioactive sources, together with associated radiation detectors, contained in logging tools, are one class of logging instruments frequently used to obtain information on the characteristics of rock formations. Amercium-241 (^{241}Am, typically 9.3 GBq (0.25 curie (Ci) to 740 GBq (20 Ci)) and cesium-137 (^{137}Cs, typically 3.7 to 110 GBq (0.1 to 3 Ci) are the radioactive materials most frequently used for this purpose.

As of December 2002, 35 companies possessed NRC licenses to use radioactive materials to perform well logging operations in the United States. There are also 211 Agreement State licensees authorized to conduct similar activities in Agreement States. Note that some well logging companies possess both an NRC license as well as an Agreement State license. Therefore, the total number of licensed well logging companies in the United States is less than the sum of all NRC and Agreement State well logging licensees.

Schlumberger is a global technology services company with corporate offices in New York, Paris, and The Hague, and has more than 80,000 employees working in nearly 100 countries. One of the business segments of Schlumberger is Schlumberger Oilfield Services. Schlumberger Technology Corporation (STC) is a wholly owned subsidiary of Schlumberger Oilfield Services. STC is authorized by NRC License 42-00090-03 to use various radioisotopes in oil, gas, mineral, coal, ground water, and environmental well logging at temporary job sites anywhere in the United States where the NRC maintains jurisdiction for regulating the use of licensed material, including areas of exclusive Federal jurisdiction within Agreement States. STC also possesses multiple Agreement State licenses to conduct similar activities in various Agreement States.

To help understand event descriptions provided in this report, the following is a brief description of the job duties of some of the workers assigned to a typical well logging crew at STC. In addition, a glossary of other terms used in the report is also included in Appendix C.

Wireline Field Engineer

This individual is responsible for ensuring that the preparation and dispatching of equipment is complete, and that the service delivered at the well site, in terms of safety, quality, and efficiency of operations, is up to standard. The wireline field engineer is in charge of his operating cell (crew) and is responsible for the training and development of personnel assigned to his cell, and for the maintenance status of his assigned equipment.

Junior Field Engineer

This individual performs tasks, as assigned by the field engineer in charge of the crew, which relate directly to the loading and unloading of radioactive sources into the logging tool. This individual also conducts radiation surveys as appropriate, and performs tasks directly related to computer processing of the data generated during wireline operations.

Senior Operator

This individual performs duties as required in the servicing of oil and gas wells and the maintenance and repair of service units and tools. Some of the many duties include (1) operating the winch for running in and out of the hole, (2) selecting, loading, and unloading required tools and corresponding surface instrumentation for the job, and (3) assisting the engineers in maintenance checks of tools and equipment. The Senior operator does not handle radioactive sources, but may be required to conduct radiation surveys after completion of logging operations, as directed by the field engineer in charge.

2 SOURCE-RELATED EQUIPMENT USED IN WELL LOGGING OPERATIONS

2.1 Overview

Many designs of well logging tools are currently in use. The tool design that is selected for a particular well logging operation depends on the subsurface geological conditions, such as well depth, heat, pressure, etc., that are present at a given well site. Once the appropriate logging tool is chosen, the radioactive source to be used in the logging operation is loaded into the logging tool. The source is loaded using a source handling tool that is designed specifically for manipulating the particular source design selected. When not in use, the well logging sources are stored and secured in shielded source transport containers. The following sections describe the source and safety-related equipment involved in the May 21st, 2002, event. Figure 2.1 shows a section of a logging tool and the hole in the tool into which the source is inserted, and Figure 2.2 shows the same tool with the source inserted in place.

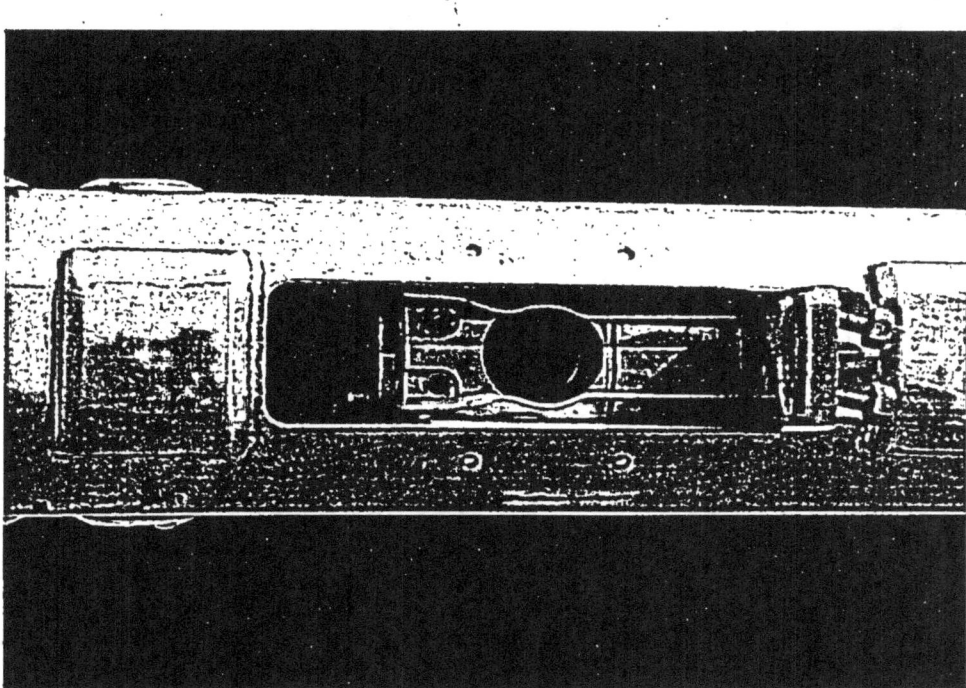

Figure 2.1 Section of the logging tool showing the slot for the radioactive source

Figure 2.2 Section of the logging tool showing the radioactive source inserted in place in the tool

2.2 Cesium-137 Source Assembly

The source assembly consists of a source capsule (Model AEA Technology X2170/2) inside a metal shield, which in turn is contained within a source housing (Model DH604538). The source capsule is a sealed source containing about 48 GBq (1.3 Ci) of ^{137}Cs. The shield that surrounds the source capsule provides substantial shielding in all directions except toward the front and to one side of the housing, where the shielding is lighter and from which the radiation is emitted for use in the logging operations. The source housing assembly is about 4.2 inches long x 1.1 inches in diameter. One end of the source assembly has a flared (dove-tail) shape that is designed to fit within the logging tool that carries the source into the well, and is also · used for picking up the source with a special source handling tool. When not in use, the source assembly fits inside a shielded storage container that is used for safe storage of the source and also to transport the source to the well sites. Figure 2.2.1 shows a photograph of the source assembly, and Figure 2.2.2 shows a detailed view of the flared end of the source assembly. Section 6.3.2 of this report provides a more complete description of the source and its radiation profile.

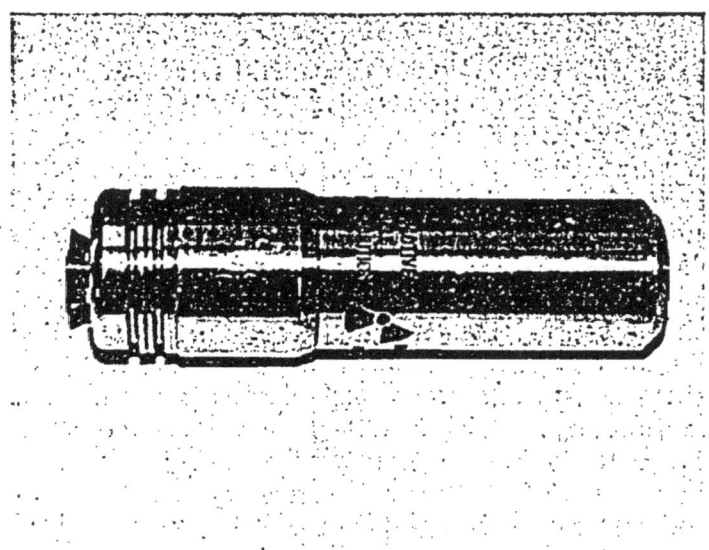

Figure 2.2.1 A cesium-137 source similar to the one used in well logging left on the rig floor during the event

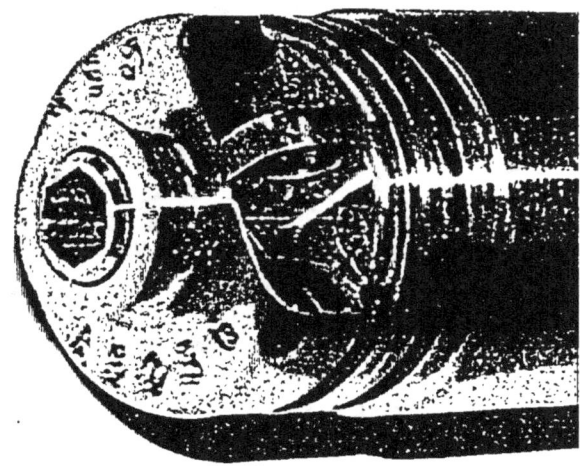

Figure 2.2.2 A detailed view of the flared end of the source assembly. The bracket, with a hole into which the safety clip is inserted, is also shown.

2.3 Source Handling Tool

The source handling tool was designed in the 1950s, and since that time, it has undergone two design changes. The first change occurred prior to around 1981 and involved (1) lengthening of the tool to 60 inches, (2) knurling of the screw-rod/turn-knob on the operator's end of the tool, and (3) removing of a flange behind the T-handle. The second design change occurred after 1987 and involved (1) redesigning and "stiffening" the outer tube of the tool, (2) introducing a "push-rod" into the annulus of the tool, (3) redesigning the safety clip so that it can be used for both the ^{137}Cs and ^{241}Am source assemblies, and (4) increasing the strength and durability of the shaft located at the grasping finger end of the tool. Figure 2.3.1 shows a photograph of the tool. The handling tool uses a grasping finger mechanism to hold onto the source assembly during transfer to and from the logging tool and the source transport container. As seen in Figure 2.2.2, one end of the source assembly has a flared (dove-tail) extension which provides a means by which the handling tool can grasp the source. Once the grasping fingers have been deployed and the source is attached to the tool, a safety-clip is then attached to the source assembly. The safety-clip is a secondary safety device to retain control of the source should it become dislodged from the grasping fingers of the tool and fall off the end of the tool during routine source handling operations. Figure 2.3.2 shows the handling tool with the source assembly and the safety clip attached. Figure 2.3.3 shows a detail of the grasping end of the tool which fits over the beveled end of the source assembly.

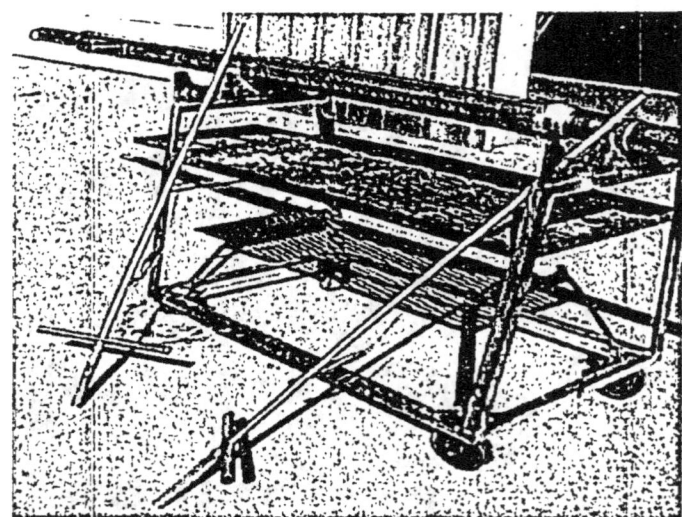

Figure 2.3.1 Two handling tools of the type used to handle the sources during the event. The end that grasps the source is at the top end of the tool in the photograph

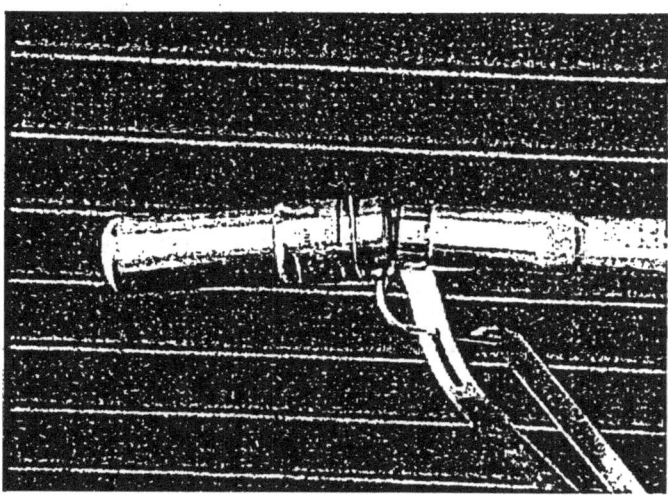

Figure 2.3.2 Source assembly attached to the end of the source tool with the safety clip shown attached to the source

Figure 2.3.3 Detail showing the grasping fingers on the end of the source handling tool. The fingers fit over the beveled end of the source assembly and are then tightened by the operator to grasp the beveled end. The knob in the center is the plunger that is used to push the source off the tool when the fingers are released.

To secure a source to the source handling tool, the field engineer places the grasping fingers at the end of the tool over the flared end of the source. He then screws the turn-knob to grasp the source with the fingers. Once the fingers have been deployed and the source is attached to the tool, the source safety clip is attached to an eyelet on the end of the source. The source can then be placed into the logging tool or source transport container as necessary. To detach the source from the handling tool, the engineer reverses the screw-rod/turn-knob to release the source, then employs the push-rod on the end of the tool to ensure that the source has been released by the tool fingers. Finally, the engineer removes the source safety clip just prior to backing the handling tool away from the source, and at the same time visually confirms that the handling tool is detached from the source.

2.4 Shielded Source Storage/Transport Container

The shielded source storage/transport container (shield), shown in Figure 2.4.1, is an STC custom-designed shielded container that weighs about 73 pounds. The container is designed such that it can be positioned straight up on its bottom or placed on its side for source loading. The container includes a shield plug insert that provides shielding above the source assembly, once the source is inserted inside the source cavity in the container. The plug insert is attached to the container with a retaining cable. When the container is used in the vertical position, as was done during the May 21st, 2002, event, the plug assembly is designed to drop into the container about 0.5 inch below the level of the container's neck, if the source assembly is not in the container. This is clearly visible, and is designed to provide a visual indication to the operator that the source is not in the container. The plug assembly will not drop in this manner when the source is in the container, or if some other factor prevents its free fall into the source well. Additionally, if the plug drops into the container, the locking mechanism for the container is designed not to be operable without having to manually raise the plug out of the source well to the proper position. This is another safety feature designed to alert the user to the absence of the source within the container. Figure 2.4.2 shows the source assembly in place within the container with the shield plug removed.

Figure 2.4.1 Top view of the source storage container with the shield plug in place and locked. The cable attaching the plug to the body of the container is also seen in the photograph

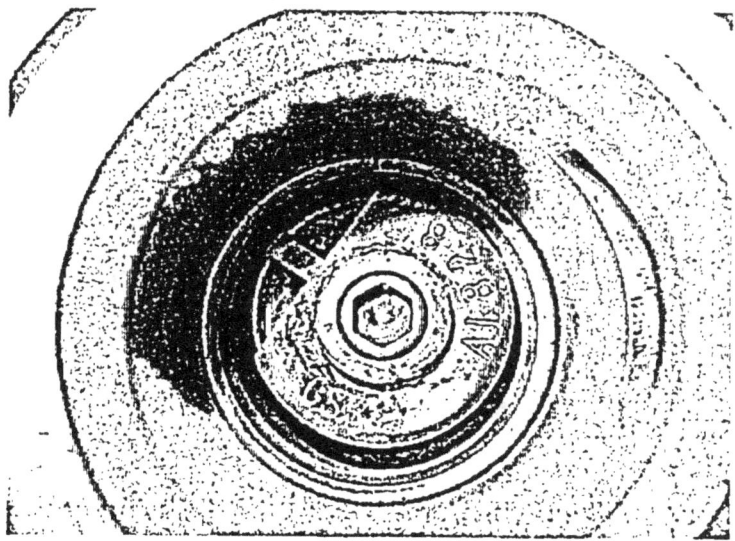

Figure 2.4.2 Top view of the source storage container with the shield plug removed showing the source assembly in place within the container storage well

9

3 SUMMARY OF EVENTS

3.1 Licensee Event Report

At approximately 16:15 p.m. (CDT) on May 23rd , 2002, the licensee's radiation safety officer (RSO) was notified by STC's Williston, ND, field office of an incident which occurred at a temporary job site located near Havre, MT, involving the loss of control of a well logging source containing approximately 48 gigabequerels (GBq) (1.3 Ci) of ^{137}Cs.   The NRC's Operations Center received notification of this event from the RSO at 16:35 p.m. (CDT) on May 23rd , 2002. The licensee reported that following the well logging operations that took place on May 21st, 2002, the logging engineers failed to properly transfer the sealed source from the well logging tool to its shielded transport container. As a result, the source was left unshielded, on the rig floor, until it was discovered missing, and subsequently recovered, two days later. A number of rig workers were believed to have been exposed to this unshielded source. These rig workers are considered to be members of the public, rather than radiation workers, because they are not exposed to radiation from licensed radioactive material as a normal part of their work. Figure 3.1.1 shows a re-enactment of the source on the rig floor in the location where it was believed to have been left after failing to be returned to the shield.

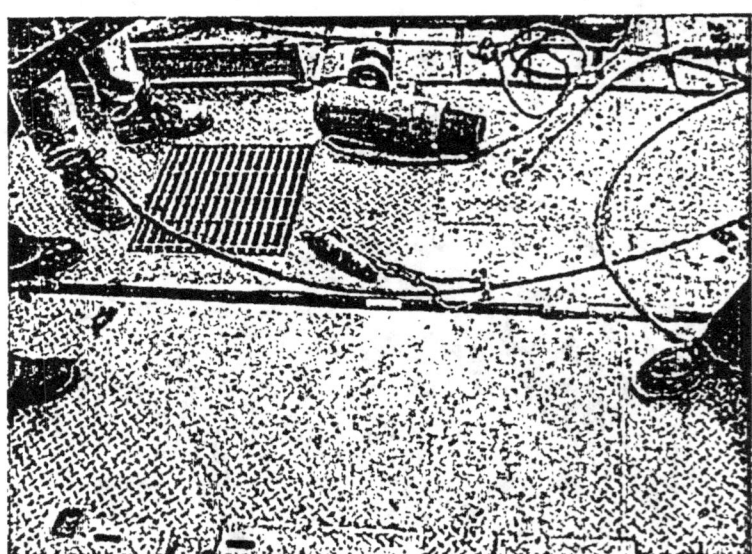

Figure 3.1.1 View of the rig platform floor where the event occurred with a dummy source shown at the location where the actual source was believed to have been left when the operator apparently failed to return it to the shielded container. The source is the small shiny cylinder in the center of the photograph.

3.2 Licensee Follow Up to the Reported Event

Following recovery of the logging source, STC sent an investigation team composed of employees from management and safety positions to Montana, beginning the evening of May 23rd, 2002. Representatives from the drilling contractor and well operator also joined the investigation. The licensee interviewed numerous individuals, including 31 rig workers believed to have been exposed to radiation as a result of the loss of control of the ^{137}Cs source.

In a written report from the licensee, dated June 25th, 2002, STC provided NRC with updated information regarding its investigation of the event. The licensee's investigation concluded that 31 members of the public (rig workers) were exposed to radiation resulting from this event. STC believed it had sufficiently bounded the exposures to these individuals and calculated the highest exposure to an individual to be approximately 6.4 centisievert (cSv) (6.4 rems). Ten workers were estimated to have received exposures between 2 and 6.4 cSv (2 to 6.4 rem), 15 individuals received exposures between 1 and 2 cSv (1 to 2 rem), and 6 individuals received exposures less than 1cSv (1 rem). As a precautionary measure, blood tests were performed on 10 of the workers to check for lymphocyte depletion, an indicator of high radiation exposure. In addition, cytogenetic testing was initiated on one of the exposed workers.

STC's June 25th, 2002, report stated that the cause for the source being left on the rig floor was determined to be a failure of the logging crew to follow standard operating procedures. Standard operating procedures required that multiple surveys with a radiation detection survey instrument be conducted to verify the presence of the source in its shielded container. None of these surveys were performed by the crew. Additionally, the licensee's investigation report stated that STC's standard operating procedures were clearly understood by the crew, that the equipment used for source transfers at the well site was properly functioning, and that STC's standard operating procedures, had they been followed, would have prevented this incident. Further, the licensee concluded that STC's training program, the equipment used for source transfers, and STC's procedures were all fundamentally sound.

The licensee's proposed corrective actions for this event included (1) terminating the employment of the three logging crew members deemed responsible for the loss of control of the ^{137}Cs source, (2) sending an "STC Alert," describing the incident, to all STC logging facilities in the United States, (3) implementing a planned modification to the licensee's training program to provide more detailed and graphic information regarding potential injuries to individuals that could occur if logging sources are not adequately secured, and (4) implementing a planned modification to the licensee's training program to include additional emphasis on the legal responsibilities of employees and managers and the potential penalties for individuals who violate company procedures.

3.3 NRC's Response to the Licensee's Event Report

In response to the licensee's telephonic report of the incident, a special, reactive inspection was initiated by NRC on May 25th, 2002, to examine the drill rig, witness interviews conducted by the licensee with some of the potentially exposed individuals, and conduct independent interviews with licensee personnel. Based on these initial interviews, and the interviews subsequently performed by the AIT, a detailed sequence of events was developed, and is included as Appendix B of this report. However, the following is a brief description of what is believed to have happened at the well site immediately after logging operations were completed on May 21st, 2002.

Using a remote handling tool, a logging engineer removed the source assembly from the well logging tool and attempted to place it in its shielded container. Assuming that the source was properly detached from the tool and properly housed in the source container, the engineer then locked the source shield plug in place. The team believes that, for reasons to be discussed later in this report, the engineer was able to lock the source plug in place despite the fact that the source was not in the container, contrary to the intended safety function of this plug, which had failed. Had this function worked properly, it would have resulted in the plug falling inside the source storage cavity, making it impossible to lock the plug in place without having to pull it out of the well, thereby alerting the engineer to the absence of the source. The AIT believes the source was probably still attached to the handling tool when the engineer pulled the tool away from the container. Then, when the engineer laid the handling tool down on the rig floor, the source probably fell off the tool and remained on the rig floor unnoticed by any of the workers. Believing that the source was safely secured in its shielded container, the logging crew then removed the source container from the rig floor and placed it in the logging truck for transport to the next job site. Figure 3.3.1 shows the location on the truck where the source container is normally stored for transportation between logging sites. The logging crew did not perform any type of radiation survey to confirm that the source was indeed shielded and secured. As a result, the ^{137}Cs source remained unshielded on the drill rig floor until it was recovered by the licensee on the evening of May 23rd, 2002, approximately 56 hours later.

After the STC logging crew left the well site, the well was completed, the portable rig was dismantled, moved to another drill site approximately 5 miles away, and there reassembled. Due to poor weather conditions, the rig remained at the new location, essentially unoccupied, for approximately 12 to 15 hours. However, during completion of the well and transfer to the new well site, some of the drilling, casing, and cementing crew members worked in close proximity of the radioactive source, possibly resulting in exposures in excess of NRC's 0.1 cSv (0.1 rem) annual dose limit to members of the public.

As indicated in Section 3.2 above, the licensee's initial assessments resulted in dose estimates for the rig workers ranging from less than 1 cSv to 6.4 cSv (1 to 6.4 rem). The AIT performed its own dose estimates for all of the exposed workers, and these showed that the licensee's estimates were high by factors of up to about 10. The AIT's dose estimates ranged from close to zero up to 1.3 cSv (1.3 rem). Blood counts and cytogenetic testing were also performed for some of the workers, and these confirmed the dose estimates by failing to show indications of high acute radiation exposures. Such negative test results indicate that, had the workers been exposed to radiation, their doses must have been less than the sensitivity of these tests. Details of the dose estimates, both those of the licensee and AIT's, as well as discussions of the blood test results, are presented in Section 6 of this report.

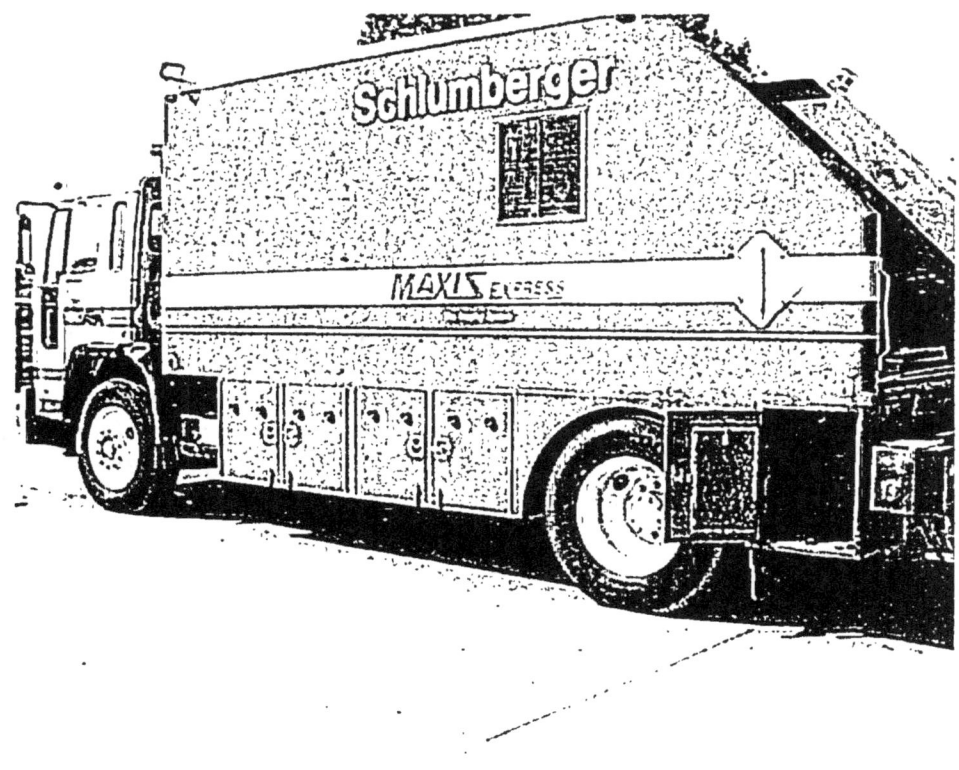

Figure 3.3.1 Well logging truck, showing the source shield storage bin, used to transport the radioactive sources between logging sites. The bin, at the bottom rear of the truck, is shown empty with its door open. The shielded source container fits into the bin.

3.4 Adequacy of the Licensee's Incident Investigation

The AIT determined that all of the exposed individuals were correctly identified by STC and that the licensee's dose estimates for these workers were reasonable based on the information available at the time of the licensee's investigation. In fact, as described in Section 6 of this report, the team concluded that STC's dose estimates were very conservative and well in excess of the doses calculated by the AIT. However, through interviews with the exposed individuals and their management representatives, the team determined that STC did not provide any follow up information regarding estimated doses, a summary of their investigation results, or information regarding the expected effects of its radiation exposures, to the exposed workers. This lack of information created anxiety for several workers who were concerned about potential health effects resulting from possible radiation exposures.

13

The team also reviewed the licensee's methodology for conducting incident investigations and STC's determination of the causal factors related to this particular event. The licensee used the "Loss Causation Model" (LCM), developed by the International Loss Control Institute, as its primary investigation tool. The individuals who conducted interviews with the exposed workers, including the RSO and others involved in the investigation, had all received specialized training in this methodology. The RSO described the LCM method as a systematic analysis tool to identify the system failures which cause incidents or "near incidents." The training slides provided to the team indicated that the fundamental principals of the LCM are (1) incidents do not just happen, (2) multiple causes usually contribute, (3) fixing immediate causes is not prevention, and basic (system) causes must be identified, (4) basic (system) causes are created by a lack of management control, and (5) plans of action must correct the basic causes and lack of control.

The team believes the LCM approach to be an adequate method for this case, and if implemented as designed, would focus the user toward discovery of contributing and root causes. Such causes, if properly addressed, can prevent similar incidents from occurring. However, the team identified weaknesses in the licensee's implementation of this method, as it was implemented following this event, as well as its use following past similar events. These weaknesses are discussed in more detail in Section 5.

4 PRECURSOR EVENTS

Through interviews with STC and other well logging licensees, the team determined that well logging sources do, periodically, fall off the handling tools during source transfers. However, these incidents are infrequent and when they do occur, the logging crews immediately identify the fact that the source has become detached, and recovery of the source is accomplished quickly. Although the number of these incidents is small in comparison to the total number of successful source transfers accomplished each day by well logging licensees, the risk of unnecessary exposures, and of exposures exceeding NRC's dose limit to members of the public, is high whenever such an event occurs. The team discovered several instances, similar to the May 2002 event, in which STC logging personnel failed to identify the fact that the sources were improperly transferred to their shielded containers, and as a result, some workers, who are considered members of the public in such cases, were unintentionally exposed to radiation.

In addition to the May 21st, 2002, event, the AIT discovered that the following six events, all involving STC personnel and equipment, occurred between 1987 and 2001 in the United States.

- In 1987, control of a 590 GBq (16 Ci) ^{241}Am neutron source was lost when, after removing the source from the logging tool, the engineer placed the handling tool, with the source still attached, on the catwalk section of the drill rig and left the site. The source remained on the job site, unshielded, for approximately 1 day. The licensee's investigation determined that a contributing cause was the failure of the engineer to perform a radiation survey of the source container or well area before leaving the site.

- In 1987, an event occurred involving the loss of control of a 63 GBq (1.7 Ci) ^{137}Cs source which was left on the rig floor, unshielded, for 4 days. The licensee's review of the incident determined that the logging engineer accidently pulled the source back out of the shield at the conclusion of the source transfer procedure. The licensee's investigation determined that a contributing cause was the failure of the engineer to perform a radiation survey of the source container or well area before leaving the site.

- In 1990, loss of control of a 63 GBq (1.7 Ci) ^{137}Cs source occurred when the source was left on the rig floor, unshielded, for an undetermined period of time. The licensee's investigation determined that the logging engineer accidently pulled the source back out of the shield at the conclusion of the source transfer procedure. The licensee's investigation also determined that a contributing cause was the failure of the engineer to perform a radiation survey of the source container or well area before leaving the site.

- In 1991, an event occurred involving the loss of control of a 63 GBq (1.7 Ci) ^{137}Cs source, which was left on the rig floor, unshielded, for 19 hours. The licensee's review of the incident determined that the logging engineer accidently pulled the source back out of the shield at the conclusion of the source transfer procedure. A contributing cause was the failure of the engineer to perform a radiation survey of the source container or well area before leaving the site.

15

- In 1997, a 63 GBq (1.7 Ci) ^{137}Cs source was knocked off the end of the handling tool when the engineer lost his balance on the rig floor. The source fell into the well bore and was not recovered.

- In August 2001, an event occurred in Edinburgh, TX involving the loss of control of a 63 GBq (1.7 Ci) ^{137}Cs source which was left unshielded on the rig floor. This incident resulted in 16 members of the public (drilling rig workers) receiving unnecessary radiation exposures, seven of which received doses in excess of 0.1 cSv (0.1 rem). The licensee's review of the incident determined that the logging engineer accidently pulled the source back out of the shield at the conclusion of the source transfer procedure. A contributing cause was the failure of the engineer to perform a radiation survey of the source container or well area before leaving the site.

The AIT requested that the licensee provide more detailed information regarding the events described above, but STC was unable to find any records that could provide sufficient information regarding the specific locations for the events, the number of potentially exposed individuals, or the estimated radiation doses resulting from these events. With the exception of the August 2001 event, dose estimates for potentially exposed individuals were not available.

5 DIRECT, CONTRIBUTING, AND ROOT CAUSES

5.1 Methods and Inspection Schedule

In selecting its methods for root cause analysis, the AIT used various analytical techniques, including Management Oversight and Risk Tree (MORT) analysis, to review human factors, equipment performance, and procedures, in an attempt to identify the casual factors related to this event. The scope of the AIT's review included (1) interviews of STC personnel and other oilfield contract personnel that were involved in the event, (2) a review of STC's sealed source handling equipment and other related sealed source safety equipment, (3) a review of STC's radiation safety training program for logging engineers, (4) a review of sealed source housing designs used by STC, as well as designs used by several other well logging companies, (5) a review of source handling tool designs used by STC as well as other designs used by several well logging companies, (6) a review of STC's inspection and maintenance program for sealed sources and safety related equipment, (7) direct observation of well logging operations at temporary job sites (oil/gas well sites) of STC and other well logging licensees, and (8) a review of STC's response to reported radiation incidents/events involving STC sealed sources, including a review of the licensee's corrective actions program.

The inspections were conducted in several Agreement States, as well as within NRC's jurisdiction. In addition, site visits were conducted at facilities in Brazil, Canada, and the United Kingdom. Table 5.1 lists the team's field site activities.

Table 5.1 Sites visited during the AIT, including those within NRC's jurisdiction, in Agreement States, Canada, Brazil, and the United Kingdom

ORGANIZATION	LOCATION
Baker Hughes INTEQ Corporate Office for U.S. Operations	Broussard, Louisiana
Century Geophysical Corporation Corporate Office	Tulsa, Oklahoma
Comprobe, Inc. Well Logging Systems Corporate Office	Fort Worth, Texas
Computalog U.S.A. Technology Services Group	Fort Worth, Texas
Halliburton Logging Services Temporary Job Site	Fairfield, Texas
Halliburton Energy Services North America Corporate Office	Houston, Texas

ORGANIZATION	LOCATION
The Henry M. Jackson Foundation for the Advancement of Military Medicine, Inc.	Rockville, Maryland
National Radiological Protection Board	Chilton, Didcot, Oxon, United Kingdom
National Nuclear Energy Com mission Institute for Radiation Protection and Dosimetry	Rio de Janeiro, Brazil
Nuclear Sources and Services, Inc. Corporate Office	Houston, Texas
Precision Drilling Technical Support Center	Nisku, Alberta Canada
Probe Technology Services, Inc Corporate Office	Fort Worth, Texas
Schlumberger Technology Corporation Temporary Job Site	Roswell, New Mexico
Schlumberger Technology Corporation Temporary Job Site	Artesia, New Mexico
Schlumberger Technology Corporation North America Corporate Office	Sugar Land, Texas
Schlumberger Technology Corporation Corporate Training Facility	Kellyville, Oklahoma
Schlumberger Technology Corporation Authorized Field Station	Williston, North Dakota
State of Texas Bureau of Radiation Control	Austin, Texas
State of Louisiana Department of Environmental Quality	Baton Rouge, Louisiana
State of Oklahoma Department of Environmental Quality	Oklahoma City, Oklahoma
Tucker Technologies, Inc. Corporate Office	Tulsa, Oklahoma
U.S. Department of Energy Oak Ridge Institute for Science and Education Radiation Emergency Assistance Center/Training Site	Oak Ridge, Tennessee
U.S. Department of Defense Armed Forces Radiobiological Research Institute	Bethesda, Maryland

ORGANIZATION	LOCATION
University of Pittsburgh University of Pittsburgh Cancer Institute Cytogenetics Facility	Pittsburgh, Pennsylvania

In addition to the locations noted above, the team conducted onsite interviews with individuals at various locations including Big Sandy, Chinook, Cut Bank, Great Falls, Kevin, Shelby and Sweetgrass, Montana; Eagan, Minnesota; Maddock and Williston, North Dakota; Baggs, Wyoming; Botha and Nisku, Alberta, Canada; and Brandon, Manitoba, Canada.

5.2 Direct Cause

The team concluded that the direct cause of the event was the failure of a logging engineer to properly transfer the ^{137}Cs source to its storage container immediately following removal of the source from the logging tool. This led directly to the loss of control event, without any additional intervening actions.

As described in Section 3.3, following logging operations, the engineer successfully removed the source from the logging tool using the appropriate source handling tool. After the engineer placed the source inside the storage well of the shielded transport container, she reversed the screw-rod/turn-knob to release the source. Believing that the source was properly detached from the tool and appropriately housed in the source storage container, the engineer then put the handling tool down on the rig floor. However, the source was apparently still loosely attached to the handling tool, and when the tool was placed on the rig floor, the source fell off the end of the tool onto the rig floor. For reasons described below, the engineer failed to recognize that the source was not properly secured in the source container. Believing the source to have been properly stored, the engineer put the shield plug in place in the source storage well and then locked the plug to the storage container. The AIT believes that the shield plug had failed at this point to drop into the source well, even though the source was not in the well. Such a drop would have made it impossible for the engineer to insert the lock into its hole on the source storage container, thereby alerting the engineer that the source may not be in place in the container.

5.3 Contributing Causes

5.3.1 Failure to Perform Appropriate Radiation Surveys

Part 39.67(c) of Title 10 of the *Code of Federal Regulations* (CFR) requires that, if a sealed source assembly is removed from the logging tool before departure from the temporary job site, the licensee shall confirm that the logging tool is free of contamination by energizing the logging tool detector or by using a survey meter. Through interviews with the responsible logging crew members, the AIT determined that this radiation survey was not performed. Although the survey is intended to identify radioactive contamination on logging tools, the team believes that the licensee's survey instrument was sufficiently sensitive to have detected the radiation emitted by the source on the rig floor. The crew would therefore have noticed elevated radiation readings and would have been alerted to an unusual condition. They would most likely have discovered the cause of the elevated readings and found the unshielded source on the rig floor.

In addition to the logging tool survey, STC's standard operating procedures also require that a "post job survey" of the source container be conducted after a logging source is locked in the container, and before the container is loaded back onto the logging truck. This requirement is intended to ensure that the source is in the container. Based on interviews with each of the three individuals on the logging crew, as well as interviews with other rig workers on site who observed the logging engineers unload the ^{137}Cs source, the team determined that a radiation survey of the source container was not performed after removing the source from the logging tool and presumably storing it in the source container, nor was a survey done prior to departure from the job site. The team's interviews disclosed that a properly calibrated radiation detection survey instrument, with the appropriate measurement range, was stored in the logging truck during the entire logging operation, but was not removed from its storage location at any time while the STC crew was at the rig site. Of equal concern to the team was the fact that two members of the logging crew signed documentation (STC's "Hazardous Material Shipping Paper - Radiation") certifying to the following statement: "I have personally checked each logging source shield with a survey meter to ensure that the source is contained within."

The AIT determined that each of the three logging crew members had received appropriate training regarding radiation surveys of logging tools and source storage containers prior to the May 2002 event. In addition, following a similar event that occurred in Texas during August 2001, each of the three individuals recalled attending a training session in which they were advised of STC's new procedure requiring that a second, independent, radiation survey of each source storage container be performed prior to the container being loaded onto the logging truck. Prior to the August 2001 event, the radiation surveys were required to be performed by only one of the logging engineers.

The team believes that if either one of the surveys described above had been properly conducted, the logging crew would have been alerted to the fact that the source was not properly shielded, which would have led to a more timely recovery of the source, thereby avoiding unnecessary exposures to workers. Therefore, the team concluded that the failure of the logging crew to perform a radiation survey of (1) the source storage container to verify that the logging source was in the properly shielded position, and (2) the logging tool immediately following removal of the sealed sources, was a contributing cause of the event.

5.3.2 False Indication by Plug Assembly

As noted in Section 2.4, the source storage container used to house the ^{137}Cs source contains, as a secondary safety feature, a plug assembly designed to provide a visual indication to the engineer when a source is not housed in the container. Specifically, if a source is not in its place in the container, then the plug assembly is designed to fall into the source storage well in the container and prevent the locking mechanism from working properly. STC's procedures specifically provide instruction to the engineers, stating, when a source is in the tool, the carrying container must remain unlocked and the container must remain vertical so that the empty container safety indicator works (lock shank cannot be inserted).

20

During interviews with STC's logging crew, RSO, and instructors at the licensee's training facility, the team was repeatedly informed that during source handling operations an engineer should recognize immediately that a source was not in the container cavity by observing that the storage container plug had dropped into the source cavity, indicating that the source was not in the container. This visual indicator is emphasized heavily during an engineer's safety training.

The AIT visited the licensee's training facility in Kellyville, Oklahoma. During this site visit, the team observed, and participated in, a simulated logging operation at STC's well logging pad used to train engineers, operators, and other STC logging personnel. The exercise included loading and unloading of "dummy sources" (source assemblies identical to those used in the field but without radioactive material in them) into logging tools and source containers. During these observations of training sessions, the team was permitted to examine and manipulate the "dummy source" and the associated safety equipment. During manipulation of the source storage container and its associated container plug, the team was able to create conditions whereby the plug assembly would not drop into the container, even when the logging source was not in the container. The team observed that when the retaining cable attached to the plug assembly became twisted, the decreased length of the cable and tension produced by the twisting did not allow the plug to fall into the container as designed. In fact, this condition allowed the plug to rest/seat in a position that was sufficient to permit locking the plug even though a source was not in the container, contrary to the intended safety function.

Figure 5.3.2.1 Photographs of the source container shield plug showing the thin, flexible, cable/clamp assembly (left) and a more rigid replacement assembly (right).

During subsequent discussions with STC personnel, and further manipulation of the source container, the team believes other conditions, besides twisting of the retaining cable, may also affect the ability of the plug to perform as designed. One such condition appears to be the size of the metal clamp used to secure the plug to the cable. The clamps observed by the team at STC's training facility and corporate offices were significantly smaller than the clamp used to

21

secure the plug to the container that was on site during the May 21st, 2002, event. Figure 5.3.2.1 shows these differences.

The team believes that the larger size clamp is more likely to get caught (i.e. "hung up") on the outer lip of the storage container, preventing the plug from dropping into the container. It appeared that various STC field locations were repairing these storage container plug cables with different size and weight clamps and cable, some of which had the potential of causing the clamp to hang up on the lip of the storage container.

In the licensee's written investigation report of the Edinburg, Texas event, STC noted that one of the corrective actions taken in response to the event was to review the design of all source shields to ensure that the plug drops when a source is not contained therein. After this review, the licensee concluded that, as designed, all of STC's source shields (storage containers) will provide this "drop." Although the plug did not drop in the Edinburg event, the licensee concluded that the failure was due to build up of drilling mud on the plug assembly; however it does not appear that STC considered other possible reasons for the container plug failure (e.g., the retainer plug cable and/or clamp size).

Based on observations of several source containers, including the one involved in the May 21st, 2002, event, and by hands-on experience manipulating the source container, the team determined that the source storage container plug may have provided a false indication to the logging engineer that the source was properly stored in the container, thereby contributing to the loss of control of the ^{137}Cs source.

5.3.3 Failure to Provide Design Specifications for Plug Assembly

The team determined that all needed repairs to both the source handling tools and source storage containers were performed by each individual STC field office, with repairs documented on records at the local STC field office. However, a record of these repairs is not provided to STC's RSO, nor is a record entered into the licensee's database for tracking equipment maintenance and repairs ("Total Rite Database"). Although the AIT did not conclude that the loss of control event was due to inadequate maintenance of associated safety equipment in general, the team believes that the failure to track the maintenance of these items could limit the licensee's ability to identify repetitive problems with a particular piece of equipment.

During interviews with the RSO and a design engineer for STC, the team learned that the primary reason for not tracking the replacement of retaining cables for the plug assembly was that the cable is not considered a safety-related device. Although the plug itself is considered a safety component, and is required to meet specific design criteria, there are no specific design criteria for the retaining cable and clamp. The team confirmed this during several site visits at which team members observed several different lengths and diameters of cables and a variety of sizes and shapes of clamps used to attach the cable to the plug. As noted in Section 5.3.2, the team determined that the ability of the plug to drop into the source container is influenced by the size and configuration of the retaining cable. This being the case, the team believes that the lack of a design specification for the retaining cable/clamp assembly directly resulted in the acquisition and use of replacement parts (cable and clamp) that may have contributed to the failure of the plug assembly to perform as designed.

In addition, although the licensee had indications from previous events that the plug assembly does not always function as designed, it does not appear that STC performed an adequate equipment failure analysis of the plug insert assembly. Specifically, aside from dirt/debris build up on the plug, the licensee did not consider other failure modes, such as the influence of the retaining cable. This limited equipment failure analysis probably also contributed to the continued practice of using replacement parts (cable and clamp) that may have been different from those in the system design, thereby defeating the intended safety function of the plug assembly, which appeared to work reliably, and as intended, only when the proper parts were used.

5.4 Root Cause(s)

The team identified as a possible root cause of the event the failure of the licensee to adequately investigate precursor events to determine their underlying causes. Instead, the licensee focused primarily on the direct cause of events and not on factors whose existence made recurrent events more probable. Section 3.4 of this report described the licensee's investigation and follow up actions taken in response to the May 21st, 2002, event. The team concluded that STC failed to execute a proper root cause analysis for this event, and likely performed a similar, limited review following other precursor events. As a result, the licensee continued to focus its corrective actions on the direct cause of events and fell short of addressing the root cause of why the errors continue to happen. Although the team agrees with the licensee that a proper radiation survey of the source container would have likely prevented the exposures to members of the public, it appears that STC's investigation(s) did not focus sufficient attention toward identifying other possible contributing causes, systems failures, and/or management controls that could prevent mishandling of sources, improper use of safety equipment, and the underlying reason(s) why logging engineers continue to fail to conduct radiation surveys and follow other standard operating procedures.

Also noted earlier in this report was that, in addition to the May 21st, 2002, event, the licensee experienced six other similar events between 1987 and 2001 involving the loss of control of logging sources. Five of these events resulted in unnecessary exposures to members of the public (unmonitored drilling rig workers). In each of these cases, STC concluded that the cause of the events was the failure of logging personnel to follow procedures. Although STC indicated that its evaluation of the May 21st, 2002, event and other similar events, included a review of people, equipment, and procedures, the team believes that these evaluations were deficient, in that the review of these areas appears to have been performed independently of each other. Specifically, it appears that (1) STC's application of its root cause analysis method did not recognize "generic" issues; causes that were identified were treated as unique (no trending or adequate root cause performed), (2) STC's application of its root cause analysis method failed to consider the contributions of its management and supervision systems on the occurrence of an event, (3) STC's corrective actions appeared to focus on the use of disciplinary action of employees for events with repetitive causes (employee error), rather than performing a more in-depth evaluation of human error, (4) the licensee's program did not contain the requirement to monitor corrective actions after they were implemented to determine if these actions were effective, and (5) STC's corrective actions program appears to rely on the use of lower tier corrective measures (procedural changes, awareness, and safety warnings) in lieu of the more effective use of safety devices and/or design changes that are higher on the "safety precedence sequence."

The team believes that the licensee's limited review of precursor events established the conditions that allowed contributing causes to develop which, in turn, increased the probability of the occurrence of future incidents.

6 RADIOLOGICAL DOSE EVALUATIONS

6.1 Overview

The licensee's initial dose calculations for the workers who may have been exposed to the well logging source on the drilling rig were preliminary estimates because details of the exposure conditions were not known in detail at that time. Thus, the methods used for the dose calculations were necessarily simplified. These calculations indicated doses that were significantly above NRC's dose limit of 0.1 cSv (0.1 rem) per year for members of the public, and were in some cases slightly above NRC's occupational dose limit of 5 cSv (5 rem) per year. Such dose levels would constitute violations of regulatory requirements, but are not considered to pose any immediate threat to the health of the workers. The dose estimates were believed by the licensee to indicate upper limits of possible doses, due to the conservative assumptions used in the calculations.

Because of the uncertainties involved in the initial calculations, and because there were no radiation measurements available at the time of the exposures, the licensee decided to supplement the preliminary dose estimates with other methods that would bound the possible doses received. These methods fall into the area known as biological dosimetry. Radiation exposures above a certain threshold level are known to produce clinically observable and other physiological effects, and the nature of these effects, as well as the time of their appearance following exposure to radiation, can be used to estimate the dose received. Absence of these biological indicators provides assurance that the doses received were at least below the thresholds at which these indicators manifest themselves.

Because biological dosimetry methods have relatively high dose thresholds, below which they do not provide indications of radiation exposure, these tests were used in this case only to rule out high doses. Accurate assessments of the doses received in this case relied on calculations which, assuming reasonably accurate input data on durations of exposure and distances from the source, will yield good dose estimates. The following sections detail the efforts made to use biological estimators of dose and describe the dose calculations made.

6.2 Biological Indicators of Dose

Acute radiation exposures (i.e., exposures in which the radiation dose is received over a relatively short period of time, usually less than 1 day) produce physiological and clinical effects if the dose is sufficiently high. The biological effects of significance in this case are prodromal symptoms, circulating blood cell depletion, and the appearance of aberrations in the chromosomes of the exposed person's circulating lymphocytes.

6.2.1 Prodromal Effects

These effects include nausea, vomiting, anorexia, fatigue, and weakness. They appear within a day or less of the radiation exposure and clear spontaneously within a day or so if the dose is not very high. The severity of these effects, and the probability of their appearance, is proportional to the whole body dose received. Experience has demonstrated that they are unlikely if the whole body dose is below a certain threshold, on the order of about 50 cGy (50 rads). None of the exposed workers in this case reported any symptoms within the first few days after exposure that may have been indicative of prodromal effects. On that basis the AIT, in consultation with REAC/TS, concluded that any doses the workers may have received were probably lower than an equivalent whole body dose of about 50 cGy (50 rads). This is consistent with the results of the licensee's preliminary dose estimates, which indicated doses well below the threshold for prodromal effects.

6.2.2 Blood Cell Depletion

Circulating blood cells are sensitive to radiation exposure, and high doses of radiation will cause a decrease in the number of these cells in circulation. Among the most sensitive of the circulating cells to radiation effects are the lymphocytes and the platelets. The drop in blood cell counts is expected to be only slight for doses in the range of 50 to 100 cGy (50 to 100 rads). As a precautionary measure, the licensee recommended that the exposed workers provide blood samples to determine circulating blood cell levels. Ten of the workers did provide such samples, and the results were sent to the Radiological Emergency Assistance Center/Training Site (REAC/TS) for evaluation. REAC/TS did not find any indications of radiation exposure based on its examination of the blood sample results for these workers. This conclusion is consistent with the results of the preliminary dose estimates, which indicated doses well below the threshold.

In addition to the one-time blood counts done for the 10 workers, one of the workers, who will be referred to here as worker D, decided on his own initiative to continue testing his blood on a weekly basis. Figure 6.2.2 shows the results of these tests, demonstrating the variations in lymphocyte and platelet counts over a period of several months. REAC/TS also reviewed these results and concluded that they were consistent with an absence of an acute radiation exposure because they do not show the typical rapid drop in blood cell counts, followed by a slow recovery, that is characteristic of acute radiation exposure. This does not mean that no radiation exposure occurred, but only that any whole body dose that may have been received was less than the level that would produce an observable blood cell and platelet count drop.

6.2.3 Cytogenetic Tests

In addition to the above effects of acute radiation exposure, ionizing radiation will also produce characteristic defects in the chromosomes of the exposed person. A sensitive method of observing these defects is by examining the chromosomes in the person's circulating lymphocytes. The defect of particular interest in quantifying radiation dose by this method is called a dicentric. This defect can be observed by microscopic examination of the cells after suitable culturing followed by microscope slide preparation and staining. Calibration curves allow conversion of the observed dicentric frequency into an equivalent whole body radiation dose.

In order to substantiate the results of the blood count tests, the licensee decided to submit a blood sample from one of the exposed workers for cytogenetic testing: a sample from one of these workers, worker D, was sent to the Armed Forces Radiobiology Research Institute (AFRRI) for analysis. This was the same worker whose blood counts are shown in Figure 6.2.2. The licensee expected that the results would be negative because the preliminary dose calculations, as well as the absence of any clinical and physiological symptoms, indicated

26

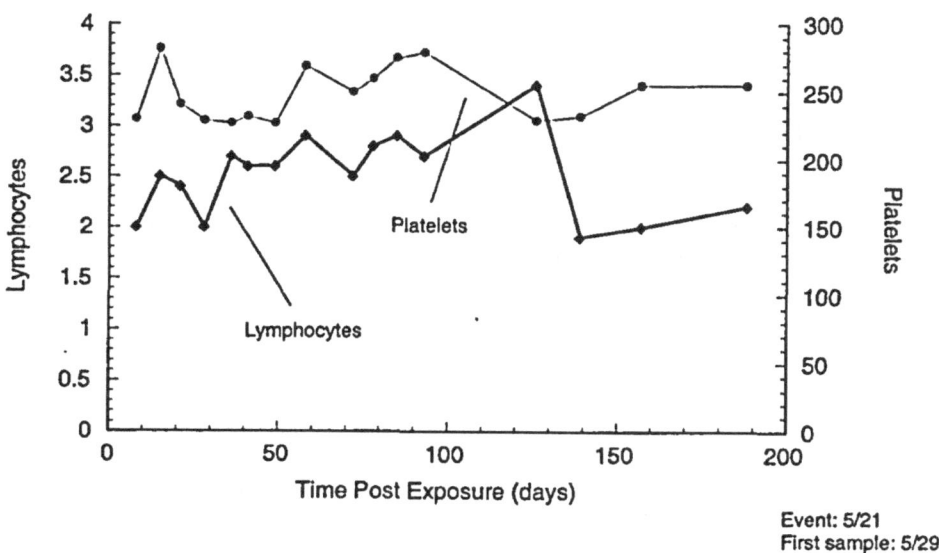

Event: 5/21
First sample: 5/29

Figure 6.2.2 Lymphocyte and platelet counts for one of the workers, Worker D, from blood samples taken at approximately weekly intervals starting about one week after the event.

that the probable doses received by any of the workers were below the sensitivity of the cytogenetic tests, which is typically about 20 cGy (20 rads). It was therefore very surprising when the cytogenetic test results indicated a likely dose of about 200 cGy (200 rads) whole body equivalent dose for this worker. By that time, NRC's detailed dose assessments had been completed, and these indicated that the most likely dose received by this worker was substantially less than 1 cGy (1 rad), or a factor of over 200 lower than the dose indicated by the cytogenetic test results. This cytogenetic result was also inconsistent with physiological evidence because a whole body dose of 200 cGy (200 rads) would be expected to produce readily observable and probably serious clinical effects and blood cell count depletion.

Because of the significant discrepancy between the calculated dose and physiological data on the one hand and the cytogenetic test result on the other, and because no immediately apparent reason could be found to explain this substantial disagreement, the AIT decided to repeat the cytogenetic test for this worker, and also to include as many of the other exposed workers as would be willing to participate. In addition, and in view of the importance of these tests, the AIT also decided to split the blood samples three ways and to send these splits to three different laboratories, including AFRRI. The three laboratories were chosen because of their recognized expertise in the field of cytogenetics, and included the *Instituto de Radioprotecao e Dosimetria*

27

(IRD) in Rio de Janeiro, Brazil, and the National Radiological Protection Board (NRPB) in Chilton, United Kingdom.

Seven workers agreed to participate in the cytogenetic testing, including worker D. In addition, two control samples were obtained, one from an individual active in the oil/gas drilling industry, but who was not present at the site of this event, and the other sample was taken from the team leader of this AIT. The blood was collected at clinics in three locations, Shelby, Montana, USA; Edmonton, Alberta, Canada; and Brandon, Manitoba, Canada. The AIT sent three sets of sampling kits to each of the three blood collection locations, and the split blood samples were packaged in the three kits at each site, together with ice packs and dosimeters provided by NRC. AFRRI provided the blood sampling kits, which contained all the medical supplies needed to draw blood samples, including disposable needles, sterile pads, sterile blood collection vials containing lithium heparin anti-coagulant, ice packs, medical release forms, and instructions. An NRC representative was present at each blood collection site to ensure proper collection, labeling, and packaging procedures. The onsite NRC representatives also shipped the packaged samples immediately by air, sending three kits to each of the three laboratories. Unfortunately, the samples shipped to Brazil were held up in Brazilian customs and could not be released in time for the blood to be analyzed before it degraded. The other two laboratories received their samples promptly and were able to analyze them.

The results of this second round of testing, as reported by the laboratory at NRPB, were negative, that is, they showed zero dose at the level of sensitivity of the tests, for all but one of the workers, worker D. This was the same worker whose blood was initially tested at AFRRI. The results from NRPB for worker D indicated a slightly elevated dicentric frequency of 2 dicentrics per 1,000 cells scored, compared with normal background, which is typically 1 dicentric per 1,000 cells scored. This level of aberration corresponds to an equivalent whole body radiation dose in the range of 0 to 14 cGy (0 to 14 rad), with a mean of 4 cGy (4 rad). The calculated dose for this worker for this event is lower than 4 cGy (4 rad), but is consistent with the cytogenetic result, which includes zero dose as a possible exposure level. The results from AFRRI agreed with those of NRPB for all the workers, that is, negative results, except for worker D, for whom AFRRI has not yet formally reported a dose estimate.

In addition to the above tests, the AIT had the microscope slides from NRPB's tests for worker D sent to IRD in Brazil for evaluation, and the slides from AFRRI's second blood test for worker D also sent to IRD for evaluation. The AIT also sent AFRRI's slides from the first blood test to NRPB for evaluation. Table 6.2.3 summarizes the results of these rounds of tests and evaluations. The table shows the results in dicentric frequency rather than dose because this makes the comparison between laboratory results more accurate by eliminating differences that may rise as a result of the use of different calibration curves by the different laboratories. The doses in cGy are very roughly equal to the numbers of dicentrics shown in the table. The results of the cytogenetics assessments by IRD and NRPB are consistent with the physiological data and also with the dose calculations. The elevated results from AFRRI are not consistent with the available data, but the reason for this is still under study at AFRRI.

Table 6.2.3 Results of the cytogenetic tests for worker D as provided by the three cytogenetics laboratories. The first blood sample was drawn in early June, and the second sample was drawn in early November

LABORATORY	DICENTRICS, NORMALIZED TO 1,000 CELLS	
	FIRST SAMPLE	SECOND SAMPLE
AFRRI	233[1]	Not reported
IRD	N/A	0[2]
NRPB	0[4]	2[3]

1. Based on 533 cells scored
2. Based on 77 cells scored. Used slides prepared by AFRRI
3. Based on 2,000 cells scored.
4. Based on 200 cells scored. Used slides prepared by AFRRI

6.3 Dose Calculations

The dose rates to the workers at the various distances from the source during this event were calculated using the Monte Carlo transport code MCNP (Monte Carlo N-Particle). This code is a numerical radiation transport code that was developed and maintained by the Los Alamos National Laboratory in New Mexico. The code allows accurate modeling of complex geometries, and has been used by NRC for many years. The workers in these calculations were represented by a mathematical model that was initially developed at Oak Ridge National Laboratory and has since been improved and updated to incorporate current information. The model contains all the important organs, tissues, and bones, and the combination of the MCNP program with the phantom permits accurate calculation of doses to each of the organs in the body in any source geometry. The calculated organ doses are weighted by the tissue weighting factors used by the NRC to obtain the effective dose equivalent, which is the quantity assessed for each worker in this case.

The Monte Carlo calculations were supplemented by two other methods, the shielding and dosimetry computer code Microshield and hand calculations using gamma ray constants and gamma ray interaction coefficients. Neither of these two methods is capable of modeling complex, inhomogeneous, geometries as is possible with MCNP. However, they are capable of producing reasonable approximations based on simplifying assumptions, and these approximate estimates were used as quality control checks on the more complex Monte Carlo calculations.

6.3.1 Time-and-Motion Study

To permit calculation of the doses received by the workers involved in this incident, the circumstances under which the exposures occurred had to be reconstructed. This involved, for each worker, estimating the distances from the source at which the worker was located during

29

Table 6.3.1 Distance and time estimates based on time-and-motion interviews with the exposed workers.

Worker Code	Minimum Distance (inches)	Maximum Distance (inches)	Time (hours)	Worker Code	Minimum Distance (inches)	Maximum Distance (inches)	Time (hours)
A	72	120	10.0	M	72	120	1.0
	36	60	2.5		36	60	2.0
B	204	240	2.0	N	72	120	4.0
	72	120	5.5		36	60	0.5
	36	60	1.5	O	72	120	5.0
C	204	240	0.5		36	60	0.5
	72	120	2.5	P	72	120	4.0
	36	60	7.0		36	60	0.5
D	204	240	4.0	Q	72	120	4.0
	72	120	1.0		36	60	0.5
	36	60	5.0	R	72	120	1.0
E	72	120	1.0		36	60	2.0
	36	60	6.0	S	204	240	4.0
F	72	120	6.0	T	204	240	3.0
	36	60	6.5		72	120	1.0
G	72	120	2.0		36	60	0.5
	36	60	2.5	U	204	240	2.0
H	204	240	4.5		72	120	5.5
	72	120	4.5		36	60	0.5
	36	60	0.5	V	204	240	2.0
K	72	120	12.0		72	120	6.0
L	72	120	4.0		36	60	0.5

Table 6.3.1 - (Continued)

Worker Code	Minimum Distance (inches)	Maximum Distance (inches)	Time (hours)	Worker Code	Minimum Distance (inches)	Maximum Distance (inches)	Time (hours)
W	204	240	2.5	BB	72	120	1.0
	36	60	1.5	CC	72	120	1.5
X	72	120	2.5		36	60	1.5
	36	60	1.5	DD	204	240	2.0
Y	72	120	1.5	EE	204	240	3.0
	36	60	1.5		72	120	5.5
Z	204	240	2.0		36	60	0.5
	72	120	1.0	FF	72	120	2.0
	36	60	0.5		36	60	3.0
AA	204	240	0.5	GG	72	120	6.5
	72	120	3.0		36	60	5.5
	36	60	1.5				

the exposure period, and the time spent at each of these distances. The licensee completed these re-enactments, which involved interviewing each worker to determine the required information. The AIT reviewed these interviews and found them to be adequate. Because of the relatively small size of the drill rig on which the exposures occurred, the distances at which the workers were exposed could be placed into the following three categories.

36 to 60 inches (90 to 150 cm)
72 to 120 inches (180 to 300 cm)
204 to 240 inches (520 to 610 cm)

The distances in centimeters were rounded to the nearest whole number. Table 6.3.1 summarizes the results of the time-and-motion studies.

6.3.2 Radiation Source

The radiation source involved in the unplanned radiation exposures was a well logging sealed source containing ^{137}Cs. The source assembly, Model AEA Technology X2170/2 Capsule, is contained in a metal shield and source holder, Model DH604538. The shield/holder serves to provide substantial shielding in all directions except toward the front and to one side of the

31

holder, where the shielding is lighter and from which the radiation is emitted for use in logging operations. The holder fits within the logging tool that carries the source into the well, and it also fits into a shielded source storage container where the source assembly is kept when not in use. The source components were mathematically modeled for use in computer calculations of the doses using the MCNP code. The source capsule and holder are stainless steel, and the shield is a tungsten alloy. The source was assayed by the manufacturer on October 29th, 1991, and determined to have an activity of 60.3 GBq (1.63 Ci). The date of the event was May 21st, 2002, and the ^{137}Cs source, with a half life of 30.0 years, would therefore have decayed to an activity of 47.4 GBq (1.28 Ci), or about 48 GBq (1.3 Ci), at the time of the event.

A source activity of 48 GBq (1.3 Ci) was used in all the dose calculations in this report, using the time-and-motion reenactment results shown in Table 6.3.1. However, because these calculations led to dose estimates that were, in at least one case, significantly different from the results of cytogenetic analysis, as discussed in Section 6.2 above, the licensee and the AIT decided to make measurements of the source strength to verify the calculated source activity. This was performed by the licensee's contractor, Nuclear Sources and Services, Inc. (NSSI), by placing a set of dosimeters at selected locations around the source to measure the dose rates at these locations. The expected dose rates were calculated by the AIT using MCNP and the mathematical model of the source and compared with the measured dose rates. The comparisons were then used to estimate a source activity. Table 6.3.2 shows the results of these measurements and calculations.

Table 6.3.2 Results of 50-centimeter circumferential measurements and axial measurements of the source, and the corresponding dose calculations. The last column shows the doses along the axis of the source at a distance of 37 centimeters

Angle	0 deg	90 deg	180 deg	270 deg	37 cm
Measured, rem/hr	1.023	0.547	0.646	0.450	2.54
Calculated, rem/hr	1.165	0.590	0.599	0.587	2.28
Ratio (Meas/Calc)	0.878	0.927	1.078	0.767	1.116

The angles for the circumferential dosimeters were measured in a clockwise direction, with zero degrees being the direction opposite the lightly shielded end of the source. The axial measurement was made at 37 centimeters from the front of the source housing along its long axis, and is shown in the last column of Table 6.3.2. The reading listed in the table at 37 centimeters is actually the mean reading of four dosimeters placed at that location. NSSI performed more measurements than those shown in the table, with the dosimeters placed at greater distances from the source than those shown. However, the results of these measurements were not used because the dosimeter readings were fairly low and the uncertainties in these readings were therefore much larger than those for the dosimeters placed closer to the source. The calculated dose rates were the dose rates calculated using MCNP with a source activity of 37 GBq (1.0 Ci). The dose calculated was the deep dose equivalent, which is the dose at a depth of 1.0 centimeter in tissue. This quantity was used because it is the quantity that dosimeters are calibrated to measure. The ratio of the measured to the calculated dose rates gives a direct indication of the source activity in Ci.

The average ratio is 0.95 ± 0.14, or a 95 percent confidence interval of 26 to 44 GBq (0.7 to 1.2 Ci). This compares well with the estimated activity of 47 GBq (1.28 Ci) based on source assay and decay correction, and the differences are probably due to uncertainties in the placement of the dosimeters, as well as uncertainties in the calibration and readings of the dosimeters. The uncertainties in the calculated dose rates are expected to be much smaller than those in the dosimeter readings. For purposes of calculating dose to workers in this event, and to ensure that doses will not be underestimated, all calculations used a source activity of 48 GBq (1.3 Ci)

6.3.3 Results of Dose Calculations

The dose rates calculated for the average and minimum source distances listed in Table 6.3.1 are shown in Table 6.3.3.1. These dose rates were used with the exposure durations shown in Table 6.3.1 for each person, and at each distance, to obtain a best estimate of that person's dose for the event. Table 6.3.3.2 details the dose estimates for each worker. Calculations show that the effective dose equivalent increases slowly as the source on the platform floor approaches the body, and then remains more or less constant at distances closer than about 50 centimeters. This behavior suggests that determination of the exact distances of the workers from the source is not critical in the calculation of a good estimate of the effective dose equivalent, and that the data obtained in the time-and-motion studies are adequate to provide reliable dose estimates. The main source of uncertainty in these calculations is probably the estimates of the times spent in the vicinity of the source, and the orientation of the body with respect to the source.

Table 6.3.3.1 - Dose rates at different distances from the source

Minimum Distance cm (in)	Average Distance cm(in)	Dose Rate mrem/hr
90(36)		130
	120 (47)	108
180 (72)		71
	240 (94)	45
520 (204)		11
	560 (220)	10

33

Table 6.3.3.2 - Dose estimates for each exposed worker based on data shown in Tables 6.3.1 and 6.3.3.1. Doses shown are effective dose equivalents

Worker Code	Dose, cSv(rem)		Worker Code	Dose, cSv(rem)		Worker Code	Dose, cSv(rem)	
	Mean	Max		Mean	Max		Mean	Max
A	0.8	1.0	N	0.3	0.4	Y	0.3	0.3
B	0.5	0.6	O	0.3	0.4	Z	0.2	0.2
C	0.9	1.1	P	0.3	0.4	AA	0.3	0.4
D	0.7	0.8	Q	0.3	0.4	BB	0.05	0.07
E	0.7	0.9	R	0.3	0.4	CC	0.3	0.3
F	1.1	1.3	S	0.04	0.04	DD	0.02	0.02
G	0.4	0.5	T	0.2	0.2	EE	0.4	0.5
H	0.3	0.4	U	0.4	0.5	FF	0.4	0.5
K	0.6	0.9	V	0.4	0.5	GG	0.9	1.2
L	0.2	0.3	W	0.2	0.3			
M	0.3	0.4	X	0.3	0.4			

6.3.4 Licensee's Dose Assessments

The licensee's preliminary dose calculations provided dose estimates for the exposed workers that are substantially higher than those shown in Table 6.3.3.2. As an example, the estimated mean and maximum doses for worker D in the table are 0.7 and 0.8 cSv (0.7 and 0.8 rem), respectively, whereas the licensee estimated a maximum dose for this worker of 4 cSv (4 rem), or a factor of about five higher than the doses shown in the table. Some of the reasons for these differences are discussed below.

- The licensee used a dose rate for the source of 74.7 millirem/hr at a distance of 115 inches, or 292 centimeters. The dose rate to tissue from an unshielded 48 GBq (1.3 Ci) ^{137}Cs point source at a distance of 292 centimeters is about 64 millirem/hr. The reason for the difference appears to be that the licensee used a source strength of 63 GBq (1.7 Ci) rather than the actual measured strength of 48 GBq (1.3 Ci). Further, the calculated dose appears to be air dose, rather than tissue dose, which is somewhat higher. These effects account for a factor of about 1.17 difference between the licensee's estimates and those in this report.

- The licensee did not appear to account for any shielding, but the source was shielded by a minimum of 0.4 centimeter of steel, which reduces the dose rate from 64 millirem/hr to about 51 millirem/hr. This accounts for an additional factor of about 1.25 difference.

- The licensee appears to have used the horizontal distance from the location of the source on the floor to the location of the worker in the dose calculations. However, the closest distance used in the calculations, which was 90 centimeters, is not representative of the distance from the source to the worker's exposed organs and tissues. For a worker of an average height of 172 centimeters, the distance from the floor to mid-torso is about 100 centimeters. The distance from the source to that point, at a floor distance from the source of 90 centimeters, is 135 centimeters, and the dose rate at that distance is a factor of about 2.25 times lower than at 90 centimeters.

- The licensee did not appear to have allowed for attenuation of the radiation in the tissue layers overlying the exposed organs. Such attenuation would be significant because of the steep angle of incidence of the radiation from the floor up toward the body. Using an average overlying tissue layer thickness of 5 centimeters, the surface dose rate is reduced by a factor of about 1.6.

Accounting for all of the above factors reduces the dose rates estimated by the licensee by an overall factor of about five, and results in closer agreement with the dose estimates shown in Table 6.3.3.2 above. The Monte Carlo calculations used in this report, based on MCNP, accurately take all of these factors into account as an integral part of the method.

Based on the above considerations, the doses in Table 6.3.3.2 will be considered to be the most accurate and reliable estimates based on physical evidence and event reconstruction. They are, however, conservative for reasons discussed in Section 6.3.5.

6.3.5 Conservatism in Dose Estimates

The dose estimates in Table 6.3.3.2 are probably upper limits on the doses likely received by the exposed workers. The conservatism in these dose estimates arises from two assumptions used in the calculations, as discussed below.

- The calculations assumed that the lightly shielded front end of the source assembly faced the workers at all times during the exposure periods. However, the dose rate from the source assembly varies markedly depending on the side from which the exposure is received, because the radiation field from the source is highly directional. For example, the dose rate from the sides of the source assembly, that is, the sides normal to the long axis of the assembly, is about 45 percent of the rate from the unshielded front side of the assembly, and is a factor of over 1000 lower when the assembly is viewed from the rear. This is due to the presence of a tungsten shield in the source assembly. Therefore, if it is assumed that the workers were exposed equally from all sides of the source during the exposure periods, the actual doses would be roughly 50 percent of the doses shown in Table 6.3.3.2.

- The calculations also assumed that the workers were exposed from the front side of the body during the entire exposure period. However, the effective dose equivalent varies

35

depending on the direction of incidence of the radiation on the body. The effective dose equivalent when the radiation is incident from the side of the body is about 50 percent of the effective dose equivalent that occurs when the radiation is incident from the front. When the radiation is incident from the back, the effective dose equivalent is about 75 percent of the dose equivalent that occurs when the radiation is incident from the front. If it is assumed that the workers were exposed equally from all sides during the exposure period, then the actual doses would be roughly 70 percent of the doses shown in Table 6.3.3.2.

Combining the two factors above, the actual doses received by the workers are probably about 35 percent of the doses shown in Table 6.3.3.2. Allowing for this factor, and using the doses at the mean distances shown in Table 6.3.3.1, the resulting estimates, shown in Table 6.3.5.1, represent the most probable doses received by the workers.

Table 6.3.5.1 Most probable effective dose equivalents received by the workers on the drill rig

Worker Code	Dose (rem)	Worker Code	Dose (rem)	Worker Code	Dose (rem)
A	0.3	N	0.1	Y	0.1
B	0.2	O	0.1	Z	0.1
C	0.3	P	0.1	AA	0.1
D	0.3	Q	0.1	BB	0.02
E	0.3	R	0.1	CC	0.1
F	0.4	S	0.02	DD	0.01
G	0.2	T	0.1	EE	0.2
H	0.1	U	0.2	FF	0.2
K	0.2	V	0.2	GG	0.3
L	0.1	W	0.1		
M	0.1	X	0.1		

6.4 Conclusions Regarding Cytogenetics Results

Table 6.2.3 shows that considerable disagreement exists in the results of the cytogenetics tests provided by the three laboratories which participated in the analysis of the blood samples for this event. The results from IRD and NRPB are in agreement, and indicate very low or zero doses for all the workers. The results from AFRRI agreed with IRD and NRPB for all the blood samples analyzed except those for worker D. Although NRPB and IRD both found very low to

36

zero doses for this worker, AFRRI found a dose of 200 cGy (200 rads) for the first sample, and did not formally report a dose for the second sample. Discussions between NRC and the laboratories involved, as well as discussions between AFRRI and NRPB, indicated that the cause of the disagreements may be an unusual characteristic in worker D's chromosomes. It appears that this characteristic causes some of worker D's chromosomes to show constrictions that, under the microscope, look like dicentrics. In addition, the arms of some of the chromosomes were crossed, again giving the impression of a dicentric when in fact none was present. Counting such constrictions and cross-overs as dicentrics would lead to erroneously high dose estimates.

Discussions were held between AFRRI and NRPB in an attempt to resolve the differences, but these discussions did not lead to agreement on the results. AFRRI's estimate still officially stands at 200 cGy (200 rad) for the first sample, and the result for the second sample has not been formally reported.

In view of these disagreements, the AIT was faced with having to decide on which of the results to accept as being most likely to be indicative of the actual exposure received by worker D. The AIT decided to accept the NRPB results for a number of reasons. In any dose assessment situation such as this one, it is essential to use all available information in arriving at the best estimate of dose, and the information that is used must be internally consistent. In this case, the available information included (1) dose calculations based on physical evidence, such as time-and-motion studies and knowledge of source activity, (2) clinical data during the period following the incident, (3) blood count data spanning the period between one week after the incident for several months subsequent to the incident, (4) reviews of the worker's clinical history, and (5) the results of the cytogenetics tests for the other six workers exposed in that incident which showed negative results. The exposure circumstances for these six workers, as determined from interviews, did not differ significantly from those for worker D. All these sources of information are consistent with a dose of the order of a centigray or less. In addition, because of the relatively low activity of the source involved, it is hard to devise a reasonable exposure scenario that would produce doses approaching the tens of centigray level, much less hundreds of centigray. Therefore, none of the known sources of information relating to this case is consistent with the AFRRI results.

7 AVAILABILITY AND TIMELINESS OF CYTOGENETIC TESTING

7.1 Process and Procedures

Cytogenetic testing to estimate radiation dose is based on the fact that ionizing radiation induces characteristic defects in the body's chromosomes, and these defects can be observed under the microscope. Several types of chromosome defects, technically called aberrations, may be used in such tests. The type used in this case, called the dicentric, is currently the most frequently used and the most reliable. For a number of reasons, the cells that are most suitable for use in such tests are circulating lymphocytes. Lymphocytes are normally not cycling, that is, they are not involved in active cell division. However, the best phase of the cell cycle to observe the chromosomes is during the metaphase part of active cell division, or mitosis. Therefore, to conduct a cytogenetics test, the exposed person's blood is drawn and special substances are added to the blood samples that induce the lymphocytes to enter active cell division. The cells are then left to incubate under strictly controlled conditions for about 48 hours, at which time many of the cells will be in metaphase. At that point, the active cycle is halted by the addition of other chemicals, and the cells are placed on glass microscope slides for examination.

The first stage of the microscopic examination is to identify the cells that were halted in metaphase. The locations of these cells on the slide are then accurately recorded for later examination. After identifying a sufficient number of metaphases, normally 200 to 1000 depending on the desired accuracy and sensitivity, the metaphases are examined carefully to identify any dicentrics that may be present. The average number of dicentrics per cell is then calculated. The photograph in Figure 7.1 shows a microscopic view of cells in metaphase, with the chromosomes clearly visible. Four, or possibly five, dicentrics are seen in this slide. The dicentrics are characterized by having two constrictions, rather than one normally seen in chromosomes at this stage of the cell cycle. The constriction actually represents a joining point of a chromosome pair, called the centromere: hence the name dicentric for chromosomes with two centromeres.

The last phase of the test is to convert the average number of dicentrics per cell into a dose. Extensive research has demonstrated that the number of dicentrics per cell is proportional to the dose, though this proportionality is not linear but usually has the form of a quadratic function. The proportionality also varies with the type of radiation to which the person was exposed, for example gamma rays, x rays, or neutrons, and it also varies with the dose rate. Therefore, each laboratory that performs cytogenetic testing usually develops a set of calibration curves, one for each type of radiation and for different dose rate ranges. The average number of dicentrics per cell is then used to enter these curves and read the dose. To do this correctly, it is necessary to know the type of radiation to which the person was exposed and the approximate dose rate. This type of information is normally easily obtained from knowledge of the incident that led to the radiation exposure.

Because of statistical uncertainties in estimating the number of dicentrics per cell, the dose estimates obtained using this method are reported as a mean dose and a 95 percent confidence interval. For example, the results of the first test for worker D were reported as 148 to 249 cGy (148 to 249 rad), which is the 95 percent confidence interval, with a mean dose of 199 cGy (199 rad). The size of the confidence interval can be reduced by examining a larger number of cells, but this process is limited because such examination is very time consuming. Much of the process of metaphase identification and dicentric counting is done manually. However, automation is sometimes used in some phases of the process, such as the identification of metaphases, using what is known as a metaphase finder. Dicentrics are almost always scored manually, however, and for that reason, cytogenetics remains to some extent an art, and its accuracy and success depend to some extent on the skill of the persons doing the work. Disagreements between laboratories analyzing samples irradiated to the same doses are therefore not uncommon, and much of the disagreements can often be traced to mis-identification of dicentrics, either by missing dicentrics or mistaking other aberrations for dicentrics.

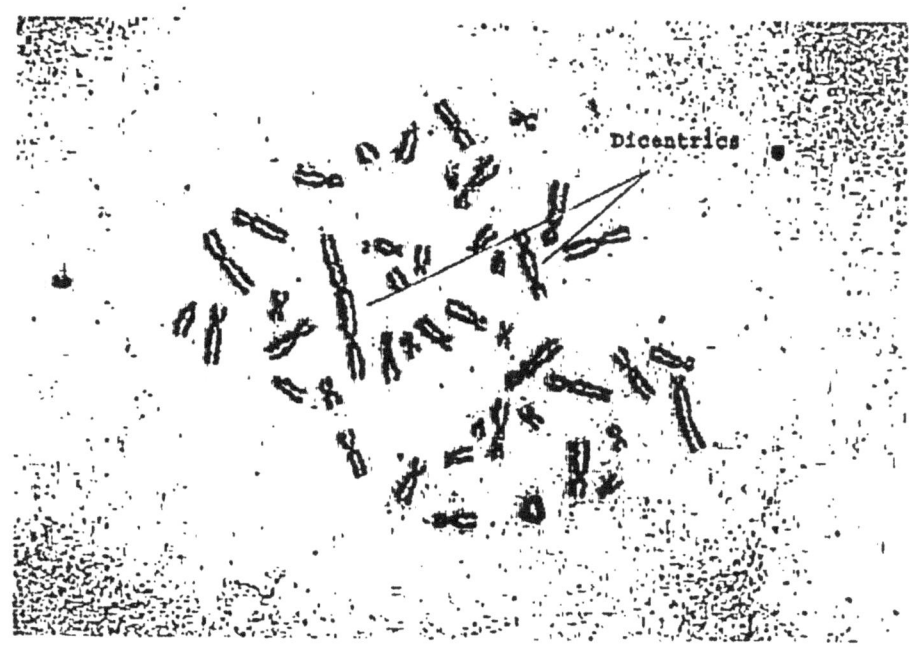

Figure 7.1 Photograph of a microscope slide showing a group of chromosomes and four, or possibly five, dicentrics, two of which are pointed out by the lines

7.2 Availability and Timeliness of Testing in the United States

The AIT identified only two facilities in the United States with laboratories currently capable of providing, biological dosimetry through cytogenetic testing, maintained the necessary calibration curves, and were prepared to perform these tests as needed. These facilities are described below:

Radiation Emergency Assistance Center/Training Site (REAC/TS), Medical and Health Science Division, Oak Ridge Associated Universities, Oak Ridge, Tennessee. The Oak Ridge Institute for Science and Education (ORISE) is a U.S. Department of Energy facility focusing on scientific initiatives to research health risks from occupational hazards, assess environmental cleanup,

39

respond to radiation emergencies, support national security and emergency preparedness, and educate the next generation of scientists. However, approximately 5 years ago, DOE ceased funding to support ORAU's cytogenetic biodosimetry program when interest shifted to lymphocyte proliferation testing for beryllium disease. NRC also subsequently eliminated its funding provided to REAC/TS for cytogenetic testing. Since that time, key personnel from the former cytogenetics staff have left the institution and are retained as consultants on an as-needed basis.

The Armed Forces Radiobiology Research Institute (AFRRI) was established in 1961, and is the sole Department of Defense research laboratory for medical radiological defense. Located in Bethesda, Maryland, AFRRI's primary mission is to develop medical countermeasures against ionizing radiation. In addition to its core objective of developing, testing, and validating deployable biodosimetry systems for military field operations, AFRRI maintains one of the nation's few reference testing facilities for radiation dose assessment. At the time of this inspection, although cytogenetic testing for civilian (non-military) accidents was outside of AFRRI's mission, the institute has provided this support on a limited, case-by-case basis, since the time that REAC/TS lost its funding.

Although several clinical laboratories throughout the United States provide cytogenetic testing services, these facilities analyze blood samples for chromosome abnormalities from the point of view of pathological conditions, such as cancer and genetic disorders. Therefore, these laboratories are not set up to perform testing for the detection of radiation induced chromosome damage. These facilities have not established the necessary calibration curves required to estimate radiation dose to an exposed individual, and lack the staffing of experienced radiation cytogeneticists.

In the past, requests for biological dosimetry have been relatively few and somewhat sporadic. As a result, many government agencies have not found sufficient justification to provide long-term funding to support laboratories whose services were needed on an intermittent basis. For these reasons, only a limited number of expert biodosimetry laboratories are currently in existence worldwide.

The current state of radiation cytogenetics testing in the United States, for use in dose assessments in unplanned acute exposures, needs improvement to assure the availability of cytogenetic testing whenever it is needed in accidental radiation exposure situations. The only facility that is now available to NRC for such testing within the United States on an ongoing and on-demand basis is AFRRI. This facility has been of assistance in the past, and may be able to assist in the future. However, AFRRI is organized and staffed as a research and armed forces support facility, and not for handling larger numbers of samples for cytogenetics testing. Cases involved more than a small number of samples probably cannot be handled to provide results rapidly enough, and in such cases it may be necessary to resort once more to overseas laboratories.

8 REGULATORY OVERSIGHT FOR WELL LOGGING LICENSEES

8.1 Regulatory Requirements

On July 14, 1987, the NRC issued a final rule amending its regulations and license requirements for the use of radioactive material in well logging. The regulation, set out in a new Part 39, consolidated radiation safety requirements for well logging into one part, established specific radiation safety requirements, and promoted the adoption of uniform radiation safety requirements among NRC and Agreement States.

The purpose of establishing a new Part 39 specifically for well logging was to have in one place in the regulations the basic safety requirements for well logging. Formerly, the requirements were often very general because they applied to many different types of licenses. The new Part 39 established specific requirements for well logging that supplemented more general requirements contained in other parts (for example, training requirements in Part 19 or survey requirements in Part 20). The Statements of Consideration for the rule note that the rule was designed to include safety requirements needed to reduce the likelihood of accidents involving the rupture of radioactive sources and the spread of radioactive contamination. Between 1982 and 1986, five accidents occurred as a result of improperly removing a stuck source from a well logging device or retrieving a well logging device lodged in a well.

Part 39.67 of Title 10 of the *Code of Federal Regulations*, "Radiation Surveys," requires, in part, that licensees (1) perform radiation surveys of the position occupied by each individual in the vehicle and the exterior of the vehicle used to transport the licensed material, (2) confirm that the logging tool is free from contamination by energizing the logging tool or by using a survey meter, and (3) make a radiation survey at the temporary job site before and after each subsurface tracer study to confirm the absence of contamination.

With the exception of the requirement to survey the position occupied by each individual in the vehicle, the survey requirements contained in Part 39 are aimed at reducing the spread of contamination. There is, however, no specific requirement in Part 39 for the licensee to survey the rig floor and source storage containers prior to leaving the temporary job site. Requirements for the conduct of surveys are, however, stated in general terms in Part 20.

Part 20.1501 of Title 10 of the *Code of Federal Regulations* requires, in part, that each licensee shall make or cause to be made, surveys that (1) may be necessary for the licensee to comply with the regulations in 10 CFR Part 20, and (2) are reasonable under the circumstances to evaluate (i) the magnitude and extent of radiation levels; (ii) concentrations or quantities of radioactive material; and (iii) the potential radiological hazards. *Survey* means an evaluation of the radiological conditions and potential hazards incident to the production, use, transfer, release, disposal, or presence of radioactive material, or other source of radiation.

Although 10 CFR 20.1501 requires that licensees perform adequate surveys to ensure compliance with the Standards for Protection Against Radiation, the team believes that a more specific requirement should be contained within 10 CFR Part 39 to require a radiation survey of the rig floor and source containers prior to leaving the site. This is consistent with the specificity of Part 39 for the specific radiological hazards of logging operations, and would also mirror similar radiation survey requirements contained in 10 CFR Part 34 (Industrial Radiography), which are aimed at providing confirmation that sealed sources have been properly return to their shielded containers at the completion of each radiographic exposure.

It should be noted that, although Part 39 does not require the surveys recommended by the AIT, STC did in fact have operating procedures in place that instruct the logging crew to perform similar radiation surveys. However, the logging crew failed to perform these surveys. As noted in Section 5 of this report, the AIT identified the failure to perform a radiation survey of the source storage container to verify that the logging source was in the properly shielded position as a contributing cause of the event.

The team also considered the need for NRC to develop a regulatory requirement for logging personnel who handle radiation sources to wear a chirper at all times during source handling operations. These are small radiation detectors that can be worn by individuals who may handle radiation sources, and provide audible signals indicating that a source of radiation is present. Chirpers and similar instruments that provide audible radiation alarms have been used by other licensees, such as radiographers, for similar purposes. The intent here is to provide well logging personnel in the field with an additional warning to indicate that a radioactive logging source is in an unshielded configuration and/or that logging personnel have failed to properly perform a radiation survey to confirm that the source has been returned to its shielded container. The team believes that the use of a radiation detection device with an audible indication that sources are unshielded may prevent an overexposure incident due to the failure of logging personnel to properly ensure the safe storage of these sources prior to leaving the well site.

Through discussions of this issue with several well logging licensees, as well as the States of Louisiana, Oklahoma, and Texas, the team determined that a more reasonable and cost effective approach to this issue is to consider the implementation of a requirement that a survey meter be physically present, with the audio feature engaged, in the immediate area where source handling is performed (rig floor, cat-walk, logging truck) at all times when sources are loaded and unloaded from the logging tools and source containers. Part 39 licensees are already required to possess operable survey instruments and to calibrate these instruments at intervals not to exceed 6 months; therefore, this requirement will present no added cost for most licensees. There may, however, be some well logging licensees who currently possess a survey instrument without an audio feature.

8.2 Licensing Program

Persons authorized by the NRC to perform well logging operations using byproduct material are licensed in accordance with the regulations in 10 CFR Part 30, "Rules of General Applicability to Domestic Licensing of Byproduct Material," and 10 CFR Part 39, "Licenses and Radiation Safety Requirements for Well Logging."

Persons authorized by an Agreement State to use byproduct material in well logging operations are licensed in accordance with corresponding statutory and administrative requirements in place in that particular State.

In addition to the regulations, the NRC has provided well logging licensees and applicants supplemental information published in NUREG 1556, Volume 14, "Program-Specific Guidance About Well Logging, Tracer, and Field Flood Study Licenses." This guide contains a compilation of information that an applicant would need to submit to be able to receive an NRC license authorizing well logging operations. The NRC staff reviews well logging license applications using, in part, the criteria published in that document.

The team found that the licensing guidance, as written, is adequate based on the current regulations. However, the guidance may need to be revised if the NRC promulgates the suggested changes to Part 39.

8.3 Inspection Program

The NRC and the Agreement States perform routine inspections of well logging licensees to determine if activities are performed in a manner that will protect the health and safety of workers and the general public and to determine if licensed programs are conducted in accordance with applicable regulatory requirements and license conditions.

NRC Inspection Manual Chapter (IMC) 2800 describes the requirements for the inspection of most of NRC's materials licenses, including well logging activities. On April 15th, 2002, NRC's Office of Nuclear Material Safety and Safeguards (NMSS) implemented Temporary Instruction 2800/033 (TI), issued to reflect a more performance-based approach to inspection activities. The TI specifies that an inspector's evaluation of a licensee's program will be based on direct observation of work activities, interviews with workers, demonstrations by workers performing tasks regulated by NRC, and independent measurements of radiological conditions at the facility, rather than reliance on a review of records. The TI also changed the normal inspection frequency for well logging licensees from 3 to 5 years. Revision 1 of the TI, issued on October 21st, 2002, specified a reduction of the inspection interval to 1 year for well logging licensees if the current inspection was limited to an office inspection and no temporary job site inspection was completed. Yet another revision to the IMC (Revision 2) was issued on December 31rd, 2002. This latest version of the TI specifies an inspection interval of 3 years (rather than 1 year as described in Revision 1) for well logging licensees if the most current inspection was limited to an office inspection and no temporary job site inspection was completed. In summary, as of December 31st, 2002, if a single inspection of a well logging licensee is conducted at a temporary job site location, the normal inspection interval for that licensee is to be changed to a 5 year frequency for inspection.

Although the TI is a "temporary instruction" aimed at collecting information to determine the appropriate course of action for the inspection program, the AIT believes that commenting on the TI is within the scope of the team's charter. Based on the expectations presented in IMC 2800, inspections are to be more performance-based in nature and are to be focused on observations of work activities. From a strictly "risk-informed" perspective, one can make the argument that 5 years, or even greater, may be acceptable. However, the team believes that direct observations of licensed activities conducted in the field by NRC inspectors is a significant indicator of licensee performance and should be conducted more often than once every five years. For this reason, the team believes that a 5-year inspection frequency for observing well logging activities is inconsistent with this "performance-based" philosophy, and should be reconsidered.

Through a review of records, and by interviews with several well logging licensees, it appears that NRC inspections rarely include the observation of licensed activities conducted at temporary job site locations. This is likely true for Agreement States as well. Some of the reasons for the difficulty in conducting field site inspections of these licensees are that the job sites are often located in very remote areas, and the logging crews are not dispatched to the site until the well reaches a specified depth, or the logging crew is on a short call-out schedule, which is often difficult to predict. This makes unannounced inspections very difficult. However, as demonstrated by team members during this inspection, "announced" inspections can be accomplished with a reasonable amount of effort, and the observations of activities in progress is worth the effort. The team encourages regional inspectors to increase their efforts to observe well logging activities in the field, even if the inspection must be announced to coordinate this effort.

It is also important to note that IMC 1220, "Processing of NRC Form 241, Report of Proposed Activities in Non-Agreement States, Areas of Exclusive Federal Jurisdiction, and Offshore Waters, and Inspection of Agreement State Licensees Operating Under 10 CFR 150.20," was revised on June 6th , 2002, and now only requires that inspections be performed for licensees with inspection priorities 1, 2, or 3. The changes imposed by TI 2800/033 (inspection priority for well logging was extended from 3 to 5 years), effectively eliminates the requirement for the regional offices to perform inspections of Agreement State licensees performing well logging activities in NRC jurisdiction. This is of concern because a significant amount of effort was expended by NRC to implement a Letter of Agreement with the U.S. Department of the Interior, Minerals Management Service, to provide NRC inspectors with helicopter transportation to temporary job sites in offshore waters, thereby allowing the agency to perform observations of well logging activities at several well site locations on an unannounced basis. The team believes that this will significantly impact NRC's initiative of increasing the agency's inspection effort at temporary job site locations in offshore Federal waters in the Gulf of Mexico. This may also have a negative impact on public confidence with regard to the regulatory oversight of licensed activities being conducted in offshore waters. Second only to industrial radiography licensees, the well logging community makes up a significant portion of licensed activities being performed in the Gulf of Mexico.

The NRC's specific procedure for performing inspections of well logging licensees are described in Inspection Procedure (IP) 87113. Section 03.06 of IP 87113, "Equipment and Instrumentation," provides guidance to the inspector, instructing the inspector to verify that the licensee has an inspection and maintenance program that provides for the visual check of source holders, logging tools, and source handling tools before each use. Additionally, the procedure states that the licensee should have an established program for the semiannual visual inspection and maintenance of source holders, handling tools, etc., to ensure that no physical damage is visible and that the required labeling is visible.

Although the inspection procedure is designed to provide the inspector with "basic guidance", the team believes that IP 87113 should be modified to provide more specific instructions regarding the minimum expectations for an inspector when reviewing a licensee's maintenance and inspection program for safety-related equipment. The team believes that inspectors should randomly examine, by visual inspection and through demonstrations by the licensee, source handling tools and source containers, to verify that the equipment functions as designed. This should be clearly articulated in the inspection procedure and/or through periodic training provided to materials inspectors.

The team's review of IP 87123, issued on December 31st, 2002, as a revised version of IP 87113, disclosed a similar concern. This revised procedure also does not provide instructions regarding the minimum expectations for an inspector when reviewing a licensee's maintenance and inspection program for safety-related equipment.

9 CONCLUSIONS

9.1 Conclusions

(a)   The team determined that well logging sources do periodically fall off the handling tools during source transfers. However, these incidents are infrequent and when they do occur, the logging crews usually immediately realize that the source has become detached and recovery of the source is accomplished quickly. Although the number of these incidents is quite small in comparison to the total number of successful source transfers accomplished each day by well logging licensees, the team discovered several instances, similar to the May 2002 event, where STC logging personnel failed to notice that sources were improperly transferred to their shielded containers, and as a result members of the public were unintentionally exposed to radiation.

(b)   The direct cause of the loss of control for this event was the failure to properly transfer the ^{137}Cs sealed source to its storage container.

(c)   Contributing and root causes of the event included (1) failure of the source storage container plug to operate correctly, which may have provided a false indication that the source was properly inserted, (2) failure to include a design specification for the retaining cable attached to the plug insert assembly for the source storage container, and failure to perform an adequate equipment failure analysis of the plug insert assembly, (3) failure to perform a radiation survey of the source storage container to verify that the logging source was in the properly shielded position, as well as failure to perform a radiation survey of the logging tool immediately following removal of the sealed sources, and (4) failure to adequately investigate precursor events to determine root causes, focusing primarily on the direct causes of events and not on factors whose existence made recurrent events more probable.

(d)   Although the team agrees with the licensee that a radiation survey of the source container would have likely prevented the exposures to members of the public, it appears that STC's investigation did not focus sufficient attention on why the source, in this and previous similar events, continued to fall off the handling tool, and why logging engineers continued to fail to conduct radiation surveys and follow other standard operating procedures.

(e)   The team determined that all of the exposed individuals were correctly identified by STC and that the licensee's dose estimates for these workers were reasonable based on the information available at the time of the licensee's investigation. As described in Section 6 of this report, the team concluded that STC's dose estimates were very conservative and well in excess of the doses calculated by the AIT. Nevertheless, through interviews with the exposed individuals, and the worker's management representatives, the team determined that STC failed to provide any followup information regarding estimated doses or a summary of its investigation result the exposed workers, even though this is required by NRC regulation. This lack of information created anxiety in several workers who were concerned about potential health effects resulting from their radiation exposure.

(f)   The NRC does not currently review or approve the design or performance standards for well logging source handling tools used by licensees to transfer sealed sources to and from their shielded containers to be loaded and unloaded into well logging tools used during logging operations. However, the team did not identify any specific generic issues or safety concerns with the design of the equipment used during logging operations.

(g)   During the course of this inspection, it became apparent to the team that numerous different designs of handling tools currently exist within the industry, and imposing a standard design would be unnecessary and impractical. This notwithstanding, in light of the number of source handling incidents, and based on interviews with several individuals, the team believes that the well logging industry should consider a further evaluation of this issue to determine if minimum engineering design and/or performance standards should be developed, similar in nature to the industry standards for associated safety equipment used in industrial radiography.

(h)   Biological dosimetry performed on some of the workers indicated doses that were below the thresholds for these methods, therefore yielding dose estimates that are below about 10 to 20 cGy (10 to 20 rad). This is generally the threshold of reliable dose estimates for the most sensitive of these methods, cytogenetic testing. These results are consistent with the dose estimates based on time-and-motion studies.

(i)   Dose estimates based on time-and-motion studies show that the doses received by any of the workers are in the range of 0 to 1 cGy (0 to 1 rad). Most of the doses are estimated to be at or slightly above NRC's dose limit for members of the public of 0.1 cSv (0.1 rem) per year, but far below NRC's dose limit for occupational exposures, which is 5 cSv (5 rem) per year.

(j)   The estimated doses are all well below the level at which clinically observable, short-term health effects would be expected. In this case, short-term here is used to mean within days to weeks of the radiation exposure.

(k)   The expected long-term effect is a slight increase in the probability of developing cancer, and possibly no effect. No direct evidence currently exists that cancer risk increases as a result of radiation exposure to doses below about 10 cGy (10 rad). However, the current approach to regulation of radiation exposure is to assume that such a risk does exist at any dose level, and that the risk increases in proportion to the dose received. The currently used risk coefficient at low doses is very small, about in 2000 per cGy of whole body dose. The cancer risk due to natural causes is about 1 in 5.

(l)   The current state of radiation cytogenetics testing in the United States, for use in dose assessments in unplanned acute exposures, urgently needs to be improved, and the appropriate Federal agencies must work collaboratively to provide a long-term solution, as well as to address the immediate need for such services.

APPENDIX A

AUGMENTED INSPECTION TEAM CHARTER

UNITED STATES
NUCLEAR REGULATORY COMMISSION
REGION IV
611 RYAN PLAZA DRIVE, SUITE 400
ARLINGTON, TEXAS 76011-8064

September 25, 2002

MEMORANDUM TO: Mark R. Shaffer, Chief
                          Nuclear Materials Inspection Branch

FROM: Ellis W. Merschoff /RA/
                          Regional Administrator

SUBJECT: **REVISED** CHARTER FOR THE AUGMENTED INSPECTION
                          TEAM FOR THE SCHLUMBERGER TECHNOLOGY
                          CORPORATION OVER-EXPOSURE EVENT (CHANGES NOTED
                          IN BOLD)

In response to recently obtained cytogenetic testing information completed for an individual exposed to radiation due to the loss of control of a Schlumberger Technology Corporation logging source, the ongoing special inspection of this event is being upgraded to an Augmented Inspection Team (AIT). You are hereby designated as the AIT leader.

A.   Basis

On May 23, 2002, Schlumberger Technology Corporation, a Region IV well logging licensee, contacted the NRC Operations Center to report the loss of control of a sealed source containing approximately 44 gigabequerels (1.2 curies) of cesium-137. A special, reactive inspection was conducted on May 25-26, 2002, to examine the drill rig, witness interviews conducted by the licensee with some of the potentially exposed individuals, and conduct independent interviews with licensee personnel.

In a written report from the licensee, received by NRC on June 26, 2002, Schlumberger Technology Corporation provided updated information regarding its investigation of the event. At that time, the licensee believed it had bounded the exposures to members of the public (drill rig crew members), and calculated the highest exposure to an individual to be approximately 6.4 rems. As a precautionary measure, blood tests were performed on 10 individuals. In addition, cytogenetic testing was performed on one of these individuals. On August 30, 2002, Region IV was provided with the results of the cytogenetic tests; the results indicate that the individual received a whole-body equivalent dose that is significantly higher than what the licensee had previously calculated. Consequently, the ongoing special inspection is being upgraded to an AIT.

Mark R. Shaffer

B.   Scope

The AIT is to examine the circumstances surrounding the exposures to members of the general public (oil/gas drilling rig crew) as a result of the loss of control of a well logging source containing cesium-137. The scope of the AIT investigation should include, but is not limited to the following:

1.   Develop a detailed sequence of events associated with the loss of control of the licensee's well logging source up to and including the results of a cytogenetic test for one individual.

2.   Determine potential root and contributing causes for the loss of control of the well logging source **including a review of any precursor events involving the licensee, and any similar events or operating experience.**

3.   Review and evaluate the scope and thoroughness of the licensee's response to and investigation of the event including the completeness and accuracy of reporting of the event.

4.   Review and evaluate the licensee's dose estimate results to determine if licensee initial estimates were reasonable based on available information.

5.   Perform independent dose estimates. Conduct interviews to determine the most likely dose scenarios.

6.   Determine if all of the potentially exposed individuals have been identified by the licensee.

7.   Evaluate the availability and timeliness of the cytogenetic testing and understand the basis for the results relative to their accuracy and validity, **including an understanding of any possible factors that may have influenced the cytogenetic results.**

8.   Review results of any additional cytogenetic testing to be performed.

9.   Identify potential generic issues.

10.   **Review the adequacy of the licensing and inspection program requirements for this type of licensee, including the need for changes.**

11.   Review the adequacy of regulatory requirements based on the risk associated with the control of well logging sources, to ensure safety and security.

C.   Guidance

This memorandum designates you as the AIT leader. Your duties will be as described in Inspection Procedure 93800, "Augmented Inspection Team." The team composition has been discussed with you directly. During performance of the augmented inspection, designated team members are separated from their normal duties and report directly to you. The team is to emphasize fact-finding in its review of the circumstances surrounding the event, and it is not the responsibility of the team to examine the regulatory process. Safety concerns identified that are not directly related to the event should be reported to the Region IV office for appropriate action.

The team will conduct an entrance meeting with the licensee at the start of the investigation, and thereafter will immediately begin a review in accordance with the scope of this charter. An exit meeting should be completed following the completion of the inspection including a review of results of any further cytogenetic studies, as deemed appropriate, with a report documenting the results of the inspection, including findings and conclusions, issued within 30 days of the exit briefing.

This Charter may be modified should the team develop significant new information that warrants review. Should you have any questions concerning this Charter, contact Dwight D. Chamberlain at (817) 860-8106.

Distribution:
M. J. Virgilio, NMSS/OD (MS 8A23)
M. V. Federline, NMSS/OD (MS 8A23)
D. A. Cool, NMSS/IMNS (MS 8F5)
R. P. Zimmerman, NSIR/OD (MS 4 D18)
E. W. Merschoff
T. P. Gwynn
D. D. Chamberlain
H. J. Miller, RI
L. A. Reyes, RII
J. E. Dyer, RIII
R. D. Hannah, PAO
W. A. Maier

APPENDIX B

SEQUENCE OF EVENTS

SEQUENCE OF EVENTS

The AIT interviewed licensee personnel and the rig workers on site during the event, reviewed licensee records and the Daily Drilling Report produced by the drilling contractor, observed the condition of the equipment involved in the event, and performed time-and-motion studies to develop the following sequence of events. Thirty-four individuals were on site during this event including 3 STC employees and 31 workers for various companies contracted for services at the well site during the event. A brief description of the type of services provided by each company is provided at the end of this section of the report.

May 20, 2002

2330   Precision Drilling completed drilling at Pimley Site Drill Rig 394 located 19 miles north of Joplin, MT to a depth of 3,498 feet, contacted STC, and STC logging crew were then dispatched to the site from Chinook, Montana to begin licensed activities.

May 21, 2002

0230-   STC (three employees) arrives at well site, designated as Pimley Site. Precision Drill Rig 394 ready for logging.

0415-   STC starts rigging up, loads sources (16 Ci ^{241}Am and 1.2 Ci ^{137}Cs) in logging tool and commences logging operations.

0600-   Harveys Water Truck Service (two employees) arrives on site.

0700-   Precision Drilling shift change; five employees on each shift. Shifts are 0700 to 1900 and 1900 to 0700.

0800-   STC completes logging operations and unloads sources from tool; the neutron source was removed first with the remote handling tool and taken to the catwalk by logging engineer 1 (LE1) where it was handed to logging engineer 2 (LE2) who then placed it in the storage container. LE2 removed the ^{137}Cs source with the remote handling tool. LE2 turned to place the source in its storage container located on the rig floor.

                [LE2 recalled placing the source in the storage container, releasing the handling tool and undoing the safety clip on the source. LE2 then placed the handling tool down on the rig floor and secured the storage container with a padlock. The storage container was removed from the rig floor by another employee.]

0800-   Harvie Hotshot crew (seven employees) arrives on site. Performs rigging operations until 1200.

                Encana representative arrives on site.
                Verploegen Excavating Service (one individual) arrives on site.

B-1

0800-0815-   Precision Drilling conducts safety meeting.

0815-1100-   Precision Drilling runs casing.

0830-1045-   STC processes well logging data: prepares for departure from site.

Harvey Water Truck Service departs site.

1045-   STC leaves well site (Pimley site), drives to Chinook, Montana where the logging truck is parked/stored, then drives back to STC's Williston, ND office in pick up truck.

1200-1205-   Precision Drilling holds safety meeting with Sanjel.

1215-1300-   Sanjel (four employees) cementing casing; departs site at 1330.

1600-   Encana representative departs site.

1700-2100-   Precision Drilling moves Rig 394 from Pimley site to Hodges site, approximately 5 miles away.

1700-2100-   Harvey Hotshot crew rigging up Rig 394 at Hodges site.

1730-   STC crew arrive Williston, ND, field office.

May 22, 2002

0000-0300-   Precision Drilling arrives at Hodges site.

Hughson Trucking arrives and remains at Hodges site for 2 hours.

Verploegen Excavating Service Truck arrives at Hodges site and remains on site for 2.5 hours.

Fugle Welding (two employees) arrived at Hodges site and remains on site for 8.5 hours.

0830-1730-   Encana representative arrives at Hodges site

1930-2200-   Encana representative arrives at Hodges site.

0300-2400-   No activity on Rig 394 due to inclement weather.

May 23, 2002

0000-2400-   No activity on Rig 394 due to inclement weather.

1400-   STC arrives at Rocky Boy well site near Havre, MT.

[During preparation for logging, LE1 notices that the ^{137}Cs source is not in its storage container. The STC crew immediately returns to the Pimley site to locate and recover the source, but discovers that Rig 394 has been moved.]

1600-   STC arrives at Hodges site: locates and recovers source off the rig floor.

May 24, 2002

0000-2400   No activity on Rig 394 due to inclement weather.

May 25, 2002

0000-2400   No activity on Rig 394 due to inclement weather.

NRC Region IV inspectors arrive on site to begin special, reactive inspection. STC conducts interviews with Precision Drilling crews and other workers on site during the event.

May 26, 2002

STC completes interviews. NRC travels to Chinook, Montana to examine source shield and associated logging equipment.

June 18, 2002

AFRRI receives one blood sample for cytogenetic testing.

July 11, 2002

AFRRI provides results of cytogenetic study to exposed individual's physician in Shelby, Montana.

August 30, 2002

Region IV receives results of the cytogenetic tests indicating a potential for other drill rig crew members to have received exposures higher than previously calculated by the licensee.

September 24, 2002

Based on results provided by AFRRI, NRC upgrades special inspection to an AIT. The following is a description of the type of services provided by the various contractors at the well site during this event.

B-3

COMPANY	JOB DESCRIPTION
Encana	Well operator
Fugle Welding	Provides welding services as needed
Harveys Water Truck	Provides water for drilling mud
Harvies Hotshot	Involved in rigging operations and moving of the drill rig
Hughson Trucking	Sets up catwalk and pipe tubs
Newpark Environmental Services	Drilling mud consultant
Noram Well Site Services	Geologist
Precision Drilling	Sets up the drill rig and performs drilling operations
Sanjel	Provides cementing service
Schlumberger Technology Services, Inc.	Provides radioactive well logging services
Verploegen Excavating	Cleans mud tanks

APPENDIX C

GLOSSARY

This glossary does not provide definitions or legal interpretations of the listed terms. Definitions are intended only to assist in reading this report.

Byproduct Material	Generally used to refer to material made radioactive as a result of nuclear reactor operation. Also includes certain uranium and thorium tailings or wastes.
Becquerel (Bq)	A unit of radioactivity equal to one disintegration per second.
Cause	The action or condition that led to the occurrence of the incident. Causes are labeled, according to their proximity to the incident, as direct, contributing, or root causes.
Curie (Ci)	A unit of radioactivity equal to 3.7×10^{10} (37 billion) disintegrations per second (Bq).
Direct cause	This is the event or failure that led directly to the incident, without any additional intervening action or failure.
Contributing cause	This is a cause that does not necessarily lead to an incident, but it does make the incident more probable.
Root cause	This is the cause whose existence establishes the conditions that allow contributing causes to develop and which, in turn, increases the probability of the occurrence of an incident.
Casing	Steel pipe placed in an oil or gas well as drilling progresses to prevent the wall of the hole from caving in during drilling and to provide a means of extracting petroleum if the well is productive.
Cementing	The application of a liquid slurry of cement and water to various points inside or outside the casing. It is used to provide a protective sheath around the casing, to segregate the producing formation, and to prevent the migration of undesirable fluids.
Mud	The liquid circulated through the well bore during rotary drilling operations. In addition to its function of bringing cuttings to the surface, the drill mud cools and lubricates the bit and drill stem, protects against blowouts by holding back subsurface pressures, and deposits a mud cake on the wall of the bore hole to prevent loss of fluids to the formations.
Wellbore	A bore hole; the hole drilled by the bit.
Radioactive well logging	The recording of the natural or induced radioactive characteristics of subsurface formations. A radioactivity log normally consists of two recorded curves, a gamma-ray curve and a neutron curve. Together

these indicate the types of rocks in the formation and the types of fluids contained in the rocks.

Trip

The operation of hoisting the drill stem from and returning it to the well bore.

Tool pusher

An employee of the drilling contractor who is in charge of the entire drilling crew and the drilling rig.

Rig up

To prepare the drilling rig for making a hole; to install tools and machinery before drilling is started.

Rig down

To dismantle the drilling rig and auxiliary equipment following completion of drilling operations.

NRC FORM 335 (2-89) NRCM 1102, 3201, 3202	U.S. NUCLEAR REGULATORY COMMISSION	1. REPORT NUMBER (Assigned by NRC, Add Vol., Supp., Rev., and Addendum Numbers, If any.)
	BIBLIOGRAPHIC DATA SHEET *(See instructions on the reverse)*	NUREG-1794

2. TITLE AND SUBTITLE

Loss of Control of Cesium-137 Well Logging Source Results in Radiation Exposures to Members of the Public

3. DATE REPORT PUBLISHED

MONTH	YEAR
October	2004

4. FIN OR GRANT NUMBER

N/A

5. AUTHOR(S)

Dennis Boal, NRC Region IV
Robert Brown, NRC Region IV
Richard Leonardi, NRC Region IV
Mark Shaffer, NRC Region IV
Sami Sherbini, NMSS, NRC Headquarters

6. TYPE OF REPORT

7. PERIOD COVERED *(Inclusive Dates)*

May 2002 to May 2003

8. PERFORMING ORGANIZATION - NAME AND ADDRESS *(If NRC, provide Division, Office or Region, U.S. Nuclear Regulatory Commission, and mailing address; if contractor, provide name and mailing address.)*

NRC Region IV

US Nuclear Regulatory Commission

611 Ryan Plaza Drive

Arlington, Texas 76011-4005

9. SPONSORING ORGANIZATION - NAME AND ADDRESS *(If NRC, type "Same as above"; if contractor, provide NRC Division, Office or Region, U.S. Nuclear Regulatory Commission, and mailing address.)*

Same as above

10. SUPPLEMENTARY NOTES

11. ABSTRACT *(200 words or less)*

This report describes the events that occurred on a drill rig in Montana on May 21, 2002, that led to the unplanned radiation exposure of 31 rig workers. These workers were not radiation workers, and were therefore considered by the Nuclear Regulatory Commission to be subject to the Agency's dose limit for members of the public, which is 0.1 centisievert (cSv) (0.1 rem) per year. The doses assessed by the Nuclear Regulatory Commission to have been received by the workers as a result of this incident were, for most of the exposed workers, above the dose limit for members of the public, but in all cases were far below the dose limit for radiation workers of 5 cSv (5 rem) per year. Although the doses received are relatively low and are not expected to cause any clinical effects, they are in violation of the Nuclear Regulatory Commission's regulations.

Included in the report is a detailed description of the sequence of events, the root and other causes, a detailed description of the methods used to assess the workers' doses, and a discussion of the biological dosimetry undertaken for some of the workers to support and verify the dose assessments. A discussion of the less than adequate state of biological dosimetry in the United States is also included.

12. KEY WORDS/DESCRIPTORS *(List words or phrases that will assist researchers in locating the report.)*

Well logging
Exposure of members of the public
Montana

13. AVAILABILITY STATEMENT

unlimited

14. SECURITY CLASSIFICATION

(This Page)

unclassified

(This Report)

unclassified

15. NUMBER OF PAGES

16. PRICE

NRC FORM 335 (2-89)

This form was electronically produced by Elite Federal Forms, Inc.

Federal Recycling Program

LOSS OF CONTROL OF CESIUM-137 WELL LOGGING SOURCE RESULTING IN
RADIATION EXPOSURES TO MEMBERS OF THE PUBLIC

UNITED STATES
NUCLEAR REGULATORY COMMISSION
WASHINGTON, DC 20555-0001

OFFICIAL BUSINESS